Biodiversity and Global Change

INTERNATIONAL UNION OF BIOLOGICAL SCIENCES

The international Union of Biological Sciences is a non-governmental, non-profit organization, established in 1919. Its objectives are to advance the development of biological sciences, to initiate, facilitate, and coordinate research and other scientific activities that require international cooperation, to encourage the discussion and dissemination of the results of cooperative research, to promote the organization of international conferences and to assist in the publication of their reports.

The membership of the IUBS presently consists of 41 Ordinary Members, adhering through Academies of Science, National Research Councils, National Research Associations, or similar organizations, and of 72 Scientific Members, all of which are international scientific associations, societies, or commissions in the various biological disciplines.

IUBS publishes *Biology International*, a news magazine, with two regular and two special issues per year, a monograph series, and special reports that appear at irregular intervals. Further information on IUBS and *Biology International* can be obtained from the Executive Director, Dr. Talal Younès, 51 Blvd. de Montmorency, 75016, Paris, France.

Biodiversity
and
Global Change

Edited by

O. T. Solbrig, H. M. van Emden

and

P. G. W. J. van Oordt

CAB INTERNATIONAL

in association with the

International Union of Biological Sciences

CAB INTERNATIONAL
Wallingford
Oxon OX10 8DE
UK

Tel: Wallingford (0491) 832111
Telex: 847964 (COMAGG G)
Telecom Gold/Dialcom: 84: CAU001
Fax: (0491) 833508

Published in association with:
International Union of Biological Sciences (IUBS)
51 boulevard de Montmorency
75016 Paris
France

A catalogue entry for this book is available from the British
Library.

ISBN 0 85198 931 4

First published 1992 by IUBS.
Reprinted with corrections and minor amendments
by CAB INTERNATIONAL 1994.

Printed and bound in the UK by Biddles Ltd, Guildford

Contents

List of Contributors

Minister *Hans Alders*
Minister of the Environment, The Netherlands

Prof. *Robert K. Colwell,*
Department of Ecology and Evolutionary Biology, University of Connecticut, Storrs, CT 06269, USA

Dr. *Patrick N. Halpin*
Environmental Sciences, University of Virginia, Charlottesville VA 22903, USA

Dr. *Andrew J. Hansen*
Forest Sciences Department, Oregon State University, Corvallis OR 97331, USA

Prof. *David Hawksworth*
International Mycological Institute, Bakeham Lane, Egham, Surrey TW20 9TY, UK

Dr. *Hiroschi Hori,*
Genetics, Gen-Iken, Hiroshima University, Kasumi, Hiroshima 734, Japan

Dr. *Otto Huber*
Instituto Venezolano de Investigaciones Cientificas, Apartado 21827, Caracas 1020-A, Venezuela

Prof. *J. W. Maurits La Riviere*
International Institute for Hydraulic and Environmental Engineering, Oude Delft 95, PO Box 3015, 2601 DA Delft, The Netherlands.

Prof. *Pierre Lasserre,*
Université de Paris VI, and Station Marine de Roscoff, 1, Place G. Teissier, 29211 Roscoff, France

Prof. *James Lovelock,*
Coombe Mill, St. Giles on the Heath, Launceston, Cornwall PL15 9RY, U.K.

Dr. *Ian A. W. Macdonald,*
Percy FitzPatrick Institute, University of Cape Town, Rondebosch 7700, South Africa. Present address: Conservation of the Southern African Nature Foundation, P O Box 456, Stellenbosch 7599, South Africa

Prof. *Robert M. May,*
Department of Zoology, Oxford University, South Parks Rd., Oxford OX1 3PS, U.K.

Prof. *Ernesto Medina,*
Instituto Venezolano de Investigaciones Cientificas, Apartado 21827, Caracas 1020-A, Venezuela

Prof. *Jean-Claude Mounolou,*
Institute Génétique et Microbol., Université de Paris-Suc, 91405 Orsay, France

Prof. *Gregoire Nicolis,*
Faculte des Sciences, Université Libre de Bruxelles, Campus Plaine, Code Postal 231, 1050 Bruxelles, Belgium.

Academician *Vladimir Sokolov,*
Institute of Evolutionary Morphology, Russian Academy of Sciences, 33, Leninskij Prospekt, 117071 Moscow, Russia

Prof. *Otto T. Solbrig,*
Department of Organismic and Evolutionary Biology, Harvard University, 22 Divinity Ave., Cambridge, MA 02138, USA

Prof. *Jan H. Stock*
Ridderweg 2, 4327 SK Serooskerke (Schowen), The Netherlands

Prof. *Dean L. Urban*
Environmental Sciences, University of Virginia, Charlottesville VA 22903 USA, currently Dept. of Range Science, Colorado State University, Fort Collins, CO 80523. USA

Prof. *Wilke van Delden,*
University of Groningen, Biologische Wetenschappen, Vg. Genetica, Postbus 14, 9750 AA Haren, The Netherlands

Prof. *Thomas van der Hammen,*
Hugo de Vries Laboratorium, University of Amsterdam, Kruislaan 318, 1098 The Netherlands

Dr. *David O. Wallin*
Forest Sciences Department, Oregon State University, Corvallis OR 97331 USA

Foreword to the Revised Edition

The IUBS symposium "Biodiversity and Global Change" held during the 24th General Assembly, 1-6 September, 1991, in Amsterdam, the Netherlands, represented the first attempt to address the issue of biodiversity from the standpoint of its ecological role, particularly, its significance in understanding global change.

Until that time, the literature dealing with biological diversity was mostly focusing on its conservation, without addressing the important questions related to the function of biodiversity, its mechanisms and change over both space and time scales. "The general lack of information and knowledge regarding biodiversity" as stated in the preamble of the Convention on Biological Diversity, adopted at the United Nations Conference on Environment and Development (UNCED), held in Rio, 992, further draws the attention to *the urgent need to develop scientific, technical and institutional capacities.*

This statement gave an additional sense of urgency to the IUBS-SCOPE-UNESCO Programme on Biological Diversity, entitled *Diversitas*. Adopted at the IUBS Amsterdam Assembly, this programme aims to fill the gaps in our knowledge of biodiversity. It addresses four themes: (1) the ecosystem function of biodiversity; (2) its origins, maintenance and loss; (3) global biodiversity inventorying and monitoring; and (4) the conservation of wild relatives of cultivated plants and domesticated animals. Also, *Diversitas* emphasises three important areas that have been neglected in the past: taxonomic resources, the diversity of microorganisms, and marine biodiversity.

The present volume, together with two other documents, *Ecosystem Function of Biological Diversity*, edited by F. di Castri and T. Younès, and *From Genes to Ecosystems: a Research Agenda for Biodiversity*, edited by O. T. Solbrig and published by IUBS in 1990 and 1991, provided the basis for the development of the conceptual framework of *Diversitas*.

A revised edition of this volume, *Biodiversity and Global Change*, was needed because of not only the wide interest shown in the development of the *Diversitas* programme itself, but also the tremendous interest shown among a wide public of scientists and teachers who are also using this volume as a reference book for university education and training concerning biodiversity.

Finally, on behalf of the International Union of Biological Sciences, I would like to pay tribute to Prof. Otto Solbrig for his leadership, dedication and incessant hard work and efforts in promoting the ideals and objectives of the Union, and also to acknowledge with thanks the contributions of Prof. van Oordt and Dr. van Emden for providing such excellent conditions for the organisation of the original symposium.

Talal Younès
Executive Director, IUBS
7 April, 1994

Preface

One of the most worrisome issues that confronts modern societies is the massive transformation of the Earth's landscapes that is taking place today, including changes in soil, water, vegetation, and atmosphere. These transformations have collectively been labelled *Global Change*.

Greatest attention so far has been paid to the climatic aspects of Global Change. Modern industrial societies have produced a large number of chemical compounds hitherto unknown on the planet, such as the family of compounds known as fluorocarbons and chlorocarbons used in refrigeration and insulation, and the chlorinated insecticides, of which DDT is the best known. Although these compounds are very effective in allowing among other things home refrigeration and air conditioning and successful control of *Anopheles* mosquitoes in the fight against malaria, they are turning to have notoriously negative side effects. In the case of the fluorocarbons it is now well established that they reduce tropospheric ozone that shields us from ultraviolet radiation, with well known and very dangerous effects on human, animal, and plant life.

The massive release of carbon dioxide resulting from the use of fossil fuels and the reduction of the standing vegetable biomass of the planet through burning and decomposition has doubled the concentration of this gas in the atmosphere. Since carbon dioxide is a so called *greenhouse gas*, that is, a gas that absorbs long wave heat energy emanating from the Earth, increases in the average temperature of the atmosphere and other climatic changes over the next hundred years have been predicted. Because climate is highly variable there still is a great deal of uncertainty regarding these predictions, contrary to the loss of tropospheric ozone which is a demonstrated certainty. Human activities have also significantly increased the concentration of methane in the atmosphere. Methane too is a greenhouse gas with effects similar to that of carbon dioxide.

Less attention has been paid to the transformation of natural landscapes with concomitant loss of natural vegetation and fauna that is taking place. Both the extent of the losses of species and their consequences for human welfare are difficult to assess and evaluate accurately. However, species loss is an irreversible process, contrary to the production of some of the human made chemical compounds, such as for example fluorocarbons, whose production is to be halted by the end of the century. Landscape transformation and species loss has been going on for several thousand of years, albeit at a slower rate. People have grown to expect landscape transformation as a natural and beneficial aspect of development. Whether such a view is tenable today has been questioned by many biologists and conservationists. Furthermore, the moral right of the human species to end the existence of other species is also being disputed.

Global change is largely the result of human activity. People influence ecosystems in a variety of ways, both direct and indirect. So, they may affect climate and the Earth's atmosphere through destruction of the ozone layer, discharge of atmospheric gases, and increase in acid rain. Or they may produce eutrophication of lakes and rivers through discharge of fertilizers or they may poison the water by dumping pesticides or other toxic substances into rivers and oceans. These and other human influences, the direct result of population increase and overconsumption are the major reasons creating global change.

This book contains the proceedings of the symposium on Biodiversity and Global Change that took place at the Royal Netherlands Academy of Sciences in Amsterdam, September 3-4, 1991 during the 24th Assembly of the International Union of Biological Sciences (IUBS). The objective of the symposium was to assess biodiversity loss in the context of global change. The book is divided into four parts.

The first three chapters are introductory dealing with organizational, political, and general questions. This is followed by four more theoretical chapters that address the meaning and practical consequences of complexity and non-linearity (Nicolis), the significance for the study of biodiversity of advances in molecular biology (Mounolou) and population genetics (van Delden), and possible interactions between biota and atmosphere (Lovelock).

The third part of the book presents a series of chapter assessing overall Biodiversity (May), in microorganisms (Hawksworth, Hori), marine systems (Laserre), various terrestrial systems (Sokolov, Medina and Huber), and in the past (van der Hammen, Stock). We close with three chapter dealing with theoretical and practical aspects of management of Biodiversity.

This symposium would not have been possible without the support and assistance of the Royal Netherlands Academy of Sciences, to whose authorities and personnel we are greatly indebted. Likewise we wish to acknowledge with thanks the financial support of the European Economic Community, the International Council of Scientific Unions (ICSU), and the United Nations Educational, Scientific and Cultural Organization (UNESCO). Professor P. G. W. J. van Oordt and Dr. H. M. van Emden served as local organizers and very gracious hosts and were responsible for the impeccable and smooth running of the sessions. We also wish to acknowledge very particularly the support of the executive committee of IUBS, and especially its past and present presidents, Professors Janis Salanki and Francesco Di Castri. Last, but not least, we acknowledge the support and encouragement, and incredible hard work of the Executive Director of IUBS, Dr. Talal Younès and IUBS' Executive Assistant, Mrs. Colleen Adams.

The editors, March 1992

1. The role of the International Council of Scientific Unions in biodiversity and global change research

J. W. Maurits La Riviere

As Secretary-General of the International Council of Scientific Unions (ICSU) I feel privileged to have the opportunity to address the Assembly of one of the largest member Unions, the International Union of Biological Sciences (IUBS) at this scientific occasion. IUBS is one of the oldest members among ICSU's 20 Unions and so is the Koninklijke Nederlandse Akademie van Wetenschappen among our 75 member Academies and research councils. ICSU is now 60 years old, and both were there at the beginning.

The title of this symposium is "Biodiversity and Global Change," and ICSU is the parent organization of the two major international research programs on global change in which biodiversity is an important element. Furthermore, ICSU has been asked by Mr. Maurice Strong, the secretary-general of the United Nations Conference on Environment and Development (UNCED), to serve as principal scientific adviser to the conference that will be held in May 1992 in Rio de Janeiro, Brazil. In this capacity ICSU has developed several channels for bringing the views of the Scientific Community officially to the attention of UNCED, and that includes their views on Biodiversity. Several participants in this symposium are already actively involved in the preparation of the Rio conference.

My presentation will be in three parts: (1) a brief explanation of what ICSU is and does; (2) a discussion of international global change research and the place of biodiversity; and (3) the "road to Rio," that is, the specific opportunities for interaction between science and policy making that UNCED provides.

The International Council of Scientific Unions (ICSU)

ICSU is an independent, nongovernmental organization. Its mandate comprises the promotion of international science; facilitating joint activities among its members; designing and implementing international interdisciplinary research programs; and acting as adviser in matters of international science. ICSU has two groups of members: the 20 Unions which are international but disciplinary, and the 75 academies which are pluridisciplinary but national. Jointly these two groups can carry out through ICSU what they could not do alone, i.e. scientific activities that are international as well as interdisciplinary.

Although they may not realize it, almost all natural scientists in the world have a relationship with ICSU, either through their national academy, or *via* an international scientific association connected to a Union, or both.

ICSU's activities can be roughly divided into two parts:
(1) activities dealing with common concerns and services, such as free circulation of scientists, the teaching of science, science and technology for development, publication of scientific results, access to data, and ethics and science; and
(2) international interdisciplinary activities, which in today's world context loom as the most important.

In its work, ICSU cooperates with some thirty associate members which include the International Institute for Applied System Analysis (IIASA) and the Third World Academy of Sciences (TWAS), as well as with the relevant United Nations organizations, in particular the United Nations Educational, Scientific and Cultural Organization (UNESCO), the World Metereological Organization (WMO), the United Nations Environmental Programme (UNEP), and the United Nations Development Programme (UNDP). ICSU has an impressive track record in international scientific programs among which figure the International Geophysical Year, the Global Atmospheric Research Programme and, *par excellence*, the International Biological Programme, in which IUBS played such an important role.

Global Change Research

Satisfied as we may be with the results of the past, we must fully realize that we are now witnessing a period of rapid transition toward a situation in which the driving forces for international interdisciplinary research are no longer mere scientific curiosities that are a push from science, but are supplemented by a strong societal imperative, a pull from Governments who urgently wish to be informed about the condition of the planet and the changes expected to occur in it, as well as suggestions for solutions.

What is understood by the study of the Earth System, or as it is often called, Global Change Research, and who are the main players? Let us emphasize that global change should not be equated with climate change, since it includes much more than that. A pragmatic definition appears to me to consist of three parts:
(1) understanding how the earth system works. The earth system comprises the three compartments land, atmosphere, water; the flows of matter cyclic and non-cyclic between them; and the biota inhabiting them. It further comprises the solar-terrestrial system to the extent it affects the earth and includes adjacent space containing satellites or their debris in orbit.
(2) finding out to what extent changes take place in the earth system, in terms of the past, the present and the future.
(3) comprehending the causes and mechanisms governing such changes.

All these studies have as a common objective the strengthening of the scientific basis for improved prediction of impacts and for the design of policy options for prevention, mitigation and adaptation measures.

The main players in Global Change Research are the International Geosphere-Biosphere Programme (IGBP) of ICSU, and the World Climate Research Programme (WCRP) conducted jointly by ICSU and the World Metereological Organization (WMO).

It is not possible in this short presentation to do justice to the wide spectrum of ICSU bodies, which includes of course IUBS, and also programs conducted outside ICSU such as the Human Dimensions of Global Environmental Change now being developed by the International Social Sciences Council, with which ICSU maintains close relations.

Before focussing on the biological aspects of these studies I would like to point out that the International Geosphere-Biosphere Programme (IGBP) and the World Climate Research Programme (WCRP) have been recognized by the United Nations General Assembly and by the Intergovernmental Panel on Climate Change (IPCC) as the major international global change research programs now in place. All nations have been invited to provide support to these programs. The execution of these programs is carried out by national research programs voluntarily integrated, or coordinated, into a coherent framework of specific projects, the design of which has in its turn been done in consultation with national committees. Networks for global change research and training are being set up in the developing parts of the world in order to make both the scientific participation and the geographic coverage of the investigation truly global: the Global Change System for Analysis, Research and Training (START).

The increasing role of biology in global change research can be illustrated by the development of the World Climate Research Programme of ICSU and WMO. Initially (1980), it focussed on the atmosphere (the fast system) and its physical aspects to which later a considerable amount of atmospheric chemistry as well as the slow part of the climate system (the oceans) had to be added. The Intergovernmental Oceanographic Commission (IOC) is about to join WMO and ICSU as a partner in WCRP. Furthermore, it became increasingly clear as these studies proceeded, that the terrestrial compartment played also an important role, as a source and sink for greenhouse gases, and that in the interaction between atmosphere with oceans as well as land, the biota plays a crucial part. Consequently biology has now become an important player in this interdisciplinary orchestra. In the planning of the brand-new Global Climate Observation System (GCOS) in *statu nascendi*, launched by the Second World Climate Conference last year, ecological expertise will be incorporated

With regard to climate change a division of work has been established between WCRP and IGBP, the former focussing on physical aspects and the latter on chemical and biological aspects. In addition, IGBP carries out biological research in areas of global change other than climate. Both groups work closely together.

Thus, we must look to IGBP for the biological component of global change research. Some of the core projects of IGBP that are relevant include:
(1) **Joint Global Ocean Flux Study** (JGOFS)
(2) **Biospheric Aspects of the Hydrological Cycle** (BAHC)
(3) Most important from the biological point of view, the huge project **Global Change and Terrestrial Ecosystems** (GCTE) with its three thrusts:

(a) Change in Ecosystem Physiology
(b) Change in Ecosystem Structure, and
(c) Global Impact on Agriculture and Forestry

All these are well established projects, underway or about to start. The following three are being considered but no final decision has as yet been made:
(4) Global Ocean Euphotic Zone Study (GOEZS) to be based on the results of the Joint Global Ocean Flux Study (JGOFS) and World Ocean Circulation Experiment (WOCE)
(5) Land Ocean Interactions in the Coastal Zone (LOICZ) which has generated great interest in the United Kingdom and the Netherlands.

Importance of Biodiversity in Global Change Studies

If we pragmatically accept that biodiversity is the "sum total of all living things on earth, taking especially into account their great variety in structure, function, and genetic make-up" we can distinguish three very different major aspects that render biodiversity one of the most important problem areas of our time:

(1) It is the living part of creation, possibly unique in the Universe;
(2) It is a storehouse of economic goods, underexplored and underexploited; and
(3) It is an essential part of the life support system. Its complexity and vulnerability are poorly understood.

The first aspect is perceived intuitively to be of great value touching the core of human existence. It is universally and powerfully embedded in culture and religion as a non-rational but no less real appreciation of the living world.

The second aspect stresses the great potential for renewable resources for food, fuel, fiber, and pharmaceuticals and other fine chemicals and thus appeals to the self-interest of the human species providing a strong argument for preservation of genetic resources and study of the 70 to 90% of the living world still unknown to us.

The third aspect in my view has drawn far less attention than it deserves. It is as yet insufficiently realized that the continued existence of humans depends on the sustained dynamic performance of the collective biota of the earth system of which we are a part. This complex biological machine together with the sun maintains the thermodynamic non-equilibrium state on earth which is the prerequisite for any form of life. It consists of many specialized interlocking parts that are attuned to one another. Each of the species or groups of species has its own structure as well as function in this global biotic machine but we do not know what happens when parts of this machine are damaged or nuts or bolts disappear. Examples of processes that may be affected are the biologically mediated fluxes between sources and sinks in the global biogeochemical cycles.

We also lack precise quantitative knowledge of these processes as evidenced by the mystery of the missing Carbon. Is it absorbed by the oceans or fixed by terrestrial vegetation? But even qualitatively our knowledge is far from complete as shown by the relatively recent discovery of the picoplankton in the Oceans, indicating that with no doubt future surprises are in store.

The importance of the latter cluster of research needs appears to have been recognized by the scientific community but still has to be grasped fully by policy makers. This brings me to the last part of this presentation: the Road to Rio

UNCED: the Road to Rio

The United Nations Conference on Environment and Development to be held in Rio de Janeiro in May of 1992 is expected to attract more than one-hundred heads of state and thousands of delegates. It is hoped that its products will include the Earth Charter, several conventions (i.e. on climate and protection of biodiversity), agreements on technology transfer, on capacity building and on financing the envisaged follow-up work which will be guided by a novel mechanisms, *Agenda 21*. This document sets out the agreed work programme of the international community. ICSU has accepted the invitation to act as principal science adviser to UNCED and in this capacity has worked out several mechanisms for bringing the contribution of the world's scientific community officially into the UNCED preparatory process. These mechanisms are all available for use by ICSU members as desired.

Among ICSU's inputs the Conference entitled Agenda of Science for Environment and Development into the 21st century (ASCEND 21, Vienna, November 1992) is the most important one, and also provided a perspective for the scientific community itself. In this effort participated some 250 natural, social, and engineering scientists from 65 countries. The conference was organized jointly by ICSU, the Third World Academy of Science, and the International Social Sciences Council.

To my knowledge, UNCED is the first occasion where the scientific community has been asked to present its views to the policy makers of the world on what should be done regarding the natural environment. Also for the first time the request asks for a very concrete answer in terms of targets, manpower and funding. If, for example, we recommend capacity building in taxonomy, we should say how many, when, and in what areas, funds are required.

I am certain that this symposium will contribute valuable ideas and recommendations for the benefit of science, of IUBS, and of UNCED.

The Road from Rio*

In the area of biodiversity the main mechanism for follow-up to UNCED is the *Convention on Biological Diversity*. Its main objectives are "... the conservation of biological diversity, the sustainable use of its components and the fair and equitable sharing of the benefits arising out of the utilisation of genetic resources". The

---

* Added in September 1993

implementation of the Convention will be kept under review by its Conference of the Parties, i.e. the countries which signed and ratified the Convention. More than 150 countries have signed and ratification is progressing. The first Conference of the Parties will take place not later than one year after its entry into force, for which ratification of at least 30 countries is required.

It is not appropriate here to discuss the contents of the Convention, which in the view of many still leaves much to be desired. But all agree that it is an important first step, providing a framework for much needed action.

As far as scientific research is concerned three articles are relevant: Article 12 calls for promotion and encouragement of research; Article 18 prescribes "International technical and scientific cooperation ... where necessary through the appropriate international and national institutions"; Article 25 establishes a "Subsidiary body on Scientific, Technical and Technological Advice" to assist the Conference of the Parties in its implementation work.

These articles provide important interfaces between the international body politic and the scientific community. In particular they provide a unique opportunity for beneficial interaction between the IUBS-SCOPE-UNESCO Programme *Diversitas* and the Conference of the Parties.

2. Towards biodiversity in politics

Hans Alders

After experiencing almost fifty years of conventional development, the world recognizes that something went wrong. Yes, conventional development has been able to raise people's living standards since the nineteen-fifties. But it also has produced world economies that have depleted stocks of ecological capital much faster than the stocks can ever be replenished.

The nature of the environmental problems for which solutions are being sought is changing. The emphasis used to lie on national and regional problems. Now it is shifting towards issues that are global and universal in character. Meanwhile there is a shift of topics from conservation of endangered species toward ill-defined issues such as Climate Change, Genetic Erosion, and Chemical Time Bombs.

Since the first United Nations Conference on Human Environment in Stockholm in 1972, considerable awareness has been raised among public and politicians, and small and big successes have been achieved. But it is likely that only a small fraction of the world community is aware of the importance of biodiversity. Most people do not realize that the functioning of local ecosystems contributes to the overall Biosphere functioning. Thus they have little apprehension that human actions, which disrupt or destroy local biological communities, can ultimately lead to loss of entire habitats and their component species. Therefore, a primary task is to raise awareness in all countries of the importance of Biodiversity.

Clearly, the existing array of protected areas and international conventions is not going to plug the species drain. Many of the biologically richest spots in the world remain unprotected. And the level of protection for existing conservation areas varies widely from site to site. Developing countries generally lack incentives to conserve biodiversity, while plenty of incentives exist to destroy biodiversity. Population and development pressures virtually ensure that natural areas will be exploited at ever increasing rates. The key in the future will be to find a way to allow these areas to be used while maintaining as high a level of biodiversity as possible.

Effects of climate change on species and plant-societies are many:

direct, through increased carbon dioxide in the atmosphere
changes in hydrological cycles
changes in the distribution patterns of vegetation
invasion of new plagues and diseases
increases in natural fires
triggering of chemical loads in soils (Chemical Time Bomb)

Increased radiation, forcing in the ultraviolet part of the spectrum, due to loss of protective ozone layer, can result in damages to chlorophyll, DNA, eye and skin pigments of all animal life and damages of general immune systems of all organisms against diseases.

One discerns three components of biodiversity: genetic, species, and ecosystem diversity. People have worried about species in recent history, but the other types of biodiversity are beyond conception of most people. And there is evolution through which biodiversity is ever changing. Geological times and regional disasters have little impact on biodiversity as a whole. But the anthropogenic forces can be disastrous in several biomes where man is abundant. The ever increasing use of land is leaving small patches of nature as islands or refugia in a biodiversity desert. In such a scene, the new evolving evolutionary forces as global warming, radiative forcing, ozone depletion, acidification and invasion of alien species or pests can be very dramatic, as there is no escape from these refugia. Neither is there any time for the organisms to adapt genetically.

What is the Situation in the Netherlands?

We have five major ecosystems in the Netherlands: coastal dunes, peat and lowlands, high sandy areas, river valleys, and tidal wetlands (Waddensea).

Current physical and land use planning at our national level stresses the linkage of big and small reserves in these five major ecosystems. This policy is the basis for our Nature Policy Plan. Of primary importance in this plan is the realization of the National Ecological Network, consisting of core areas, nature development areas, and ecological corridors. In the first period of this plan (covering eight years) many projects and actions will be realized, costing up to 155 million Dutch guilders in 1994. The planning is coordinated as much as possible with our neighbors Germany and Belgium.

We are now working on a project on common ecotypology for environmental and land use planning. The project is coordinated by four institutes: the National Institute for Health and Environment (RIVM), the National Institute for Nature Conservation (RIN), Leiden University Environmental Center and the Wageningen University Center for Climate Studies.

Not only in the Netherlands, but also in the rest of the world the problems of biodiversity should be dealt with. We are about to live in an interdependent world, in which the common cause hopefully prevails over national interests. The upcoming United Nations Conference on Environment and Development offers a unique opportunity to make further progress in international decision making in several environmental issues, including a convention on biological diversity. In preparation, many meetings are scheduled but progress is slow. Part of this discussion focuses on technology transfer towards developing nations, in order to limit environmental degradation resulting from their legitimate economic growth. Here we can learn from the negotiations of the Montreal Protocol. This Protocol of 1987 has set a new trend in motion, namely making available additional resources to developing nations to secure their participation. It is hoped that this approach can also be successfully implemented in future treaties such as on climate change and biological diversity.

In times of disputes, concerned parties should take their cases to a court for international environmental law, as I argued at last month's International Environmental Law Conference in The Hague.

Within development cooperation the Netherlands government has made it compulsory, from 1991 onward, that in concerned development projects environmental criteria will be used in project appraisal.

In international fora the Netherlands government stresses the importance of *in-situ* conservation. In the ultimate situation, *ex-situ* conservation can only be regarded as complementary. Nowhere and never, the natural environment of an endangered species can be mimicked.

Our thinking and acting should be based on the precautionary principle and aimed at concrete commitments. Each year that we continue "business as usual" policies, we reduce our own flexibility to make the required policy responses. Moreover, future costs of counter acting measures will become much more expensive.

Needed International Actions

Governments have reached during recent years remarkable international consensus on actions against carbon dioxide and CFC-emissions, as well as toxic wastes and trade in toxic chemicals. It was not easy to reach consensus on these issues. However biodiversity presents even more complex problems.

Biologists in the field are the proper judges of processes that diminish biodiversity and of changes in the ecosystem and their functions. But they do not only have to look at these processes. They must come out of their privileged "hides" into the open field and provide suggestions to the ignorant public and to politicians. Projections on timescales more than 25 years ahead are warranted. Particularly on future changes that may have serious long-term repercussions.

Biology is also the correct discipline to point out that politicians cannot contend themselves with a random starting engineering projects such as forestry and de-canalization. The quality of natural ecosystems cannot be brought back by putting some trees together, by creating a new riverbed or restricting the use of fertilizers. On the contrary, clear signals have to come out of why and how industrial societies have to restrict themselves in their use of available natural resources. Ecological field research is often restricted to small-scale ecosystems, preferably the more natural ones. For supranational politics, one needs clear ecological conceptions from the local ecosystem integrated to wider geographical scales.

The need to have more biologists and ecologists partaking in policy development in different fora, is evident. The uniqueness of an ecologist trained in field and laboratory, with broad insights into the structure and function of ecosystems, fluxes, sources and sinks, variability and balances, may shed new light on global changes in the environment.

Politics is also in need for clear suggestions on what is to come first on the list of environmental problems. This may differ from country to country, but undoubtedly biodiversity scores high anywhere. To solve these global environmental problems, we will need the help of the International Union of Biological Sciences and the ecological community at large, in outlining regional networks and create greater resilience to global change.

The world situation today, more than in 1972, makes it possible to take further concrete steps to jointly handle transboundary environmental problems. But we do need additional financial resources to find sustainable solutions. We also need the knowledge and help of biologists to tackle global change issues.

3. Biodiversity: An introduction

Otto T. Solbrig

Introduction

Biodiversity is the property of living systems of being distinct, that is, different, unlike. The word is a contraction of Biological Diversity, i.e. the diversity of living beings. Life comes in an almost infinite variety of fascinating and enchanting forms, from microscopically small unicellular species to giant whales and elephants. In turn, species are formed by different kinds of populations, these by different kinds of individuals, and these by different types of organs, tissues, cells, and genes. Diversity surround us, engulfs us, and not only in the living world. The inanimate world is also highly diverse. The rare, the peculiar, is to encounter living beings that are identical. We wonder at twins, because it is so unusual for two human beings to be so similar.

Biodiversity is therefore not an entity, a resource, but a property, a characteristic of nature. Species, populations, certain kinds of tissues are resources, but not their diversity as such. But diversity is a defining characteristic of life. Without diversity life is not conceivable, just like a ball without roundness is not conceivable, or a diamond without hardness is not a diamond. There is so much diversity, and it takes so many forms that we ask in disbelief, where does all this diversity come from? Fundamentally it derives from the properties of a variety of macromolecules, most notably DNA and proteins. Their characteristics make biodiversity possible.

Origin of Biodiversity

The DNA molecule like any other molecule must obey all physico-chemical laws. A fundamental rule is that it will adopt the state of greatest free energy, or highest entropy, following the second law of thermodynamics. The DNA molecule is a polymer formed by two long parallel backbone chains of ribose sugars linked together by phosphodiester bonds. From each sugar protrude one of four distinct organic bases that are linked to a base in the opposite chain. The DNA molecule reaches the state of greatest free energy when it forms the well known double helix, when the adenines in one chain are lined up with thymines in the opposite chain, and all glycines line up with cytosines. Yet it does not matter in which proportion or in which order the A-T and G-C pairs occur in the polymer. But although the order and proportion of the base pairs does not matter from the physico-chemical point of view they matter a great deal from the biological point of view, because the order of the A-T and G-C pairs determines the order of the amino-acids in proteins, and ultimately all the characteristics of organisms. Strings of 3 letter words with suitable adjuncts form a gene. Each gene codes for a protein, and it takes many proteins to make even the simplest of organisms. It is this combination of things, on one side a strict rule regarding the pairing of Adenine with Thymine and Glycine with Cytosine, with no rule regarding the order in which they occur along the chain that makes life possible and mandates that it be diverse. All this tremendous and marvellous diversity of life emanates from this simple duality in the DNA molecule! Given the characteristics of DNA life without diversity is impossible.

Since the order of the A-T and G-C's does not matter from a physico-chemical point of view, one base pair can be substituted for another without affecting the thermodynamic stability of the molecule. Such changes occur routinely due to a variety of causes. We call such alterations *mutations*, which means changes in Latin. On the average there is one mutation per cell division. But a shift in the order of the bases will affect the genetic code. Such a change can affect the characteristics of the protein that the modified gene makes. This is not inevitable because the code is redundant (different triplets code for the same amino-acid), and proteins can often sustain changes in the order of their amino-acid without impairing their function. A change in the order of the amino-acids can be either:

(a) **Neutral** the resulting protein has not changed functionally;
(b) **Lethal or semilethal** the resulting protein is not functional, or is deficient;
(c) **Improved** the resulting protein is more efficient.

The first type of mutations can accumulate and create diversity that apparently does not affect function. The extent and importance of this neutral variation has been the subject of much debate (Crow and Kimura 1970).

The second kind of mutations are eliminated by natural selection. Lethal and semilethal mutations have been the target of a great deal of research in genetics. Among the research subjects are the frequency of occurrence and the mechanism of elimination of mutations.

The last type of mutation is the material that ultimately gives rise to all biological diversity. There is also a great deal of debate regarding the role of these mutations. Does evolution proceed by the gradual accumulation of small mutations, or are there occasional *megamutations*, changes so large and important that they can give rise to new species and lineages, as claimed by some researchers (Goldschmidt 1945). This is a subject that is hotly debated these days. A very disputed question is whether mutations direct evolution by the so called processes of *mutation pressure* (Lima-de-Faria 1988) or whether, as advocated by the Neodarwinian school of evolution (Mayr 1963) there is enough underlying variation as a result of mutations that evolution is to be seen primarily in terms of natural selection.

Biological Diversity

Variation is a fundamental property of life. There is actually so much genetic variation that the numbers of variants are beyond our ordinary experience. For example, even in asexually reproducing species, individuals differ by at least one mutation, so that the number of genetically different organisms at any one time is slightly less than the number of organisms alive. It is impossible to estimate that accurately. The number of individuals of a species varies tremendously both in time and from species to species. Some, such as our own, count themselves in the thousands of millions, while others, have fewer than 100 individuals. These are the species in danger of extinction, such as the greater panda, the California Condor, or the whooping crane. If I had to put forth a wild guess as to what the average number of individuals per species is, I would say it is over 10 000 [10^4], in my opinion a very conservative number. As to the number of species, there are so many kinds, that taxonomists in over 200 years of research have not yet been able to estimate precisely how many there are. A conservative number is 10 million [10^7] (May 1988, this book), although estimates as high as 100 million have been made (Ehrlich and Wilson 1991). Consequently there are, conservatively speaking, a very minimum of 10^{11} [10^4 x 10^7] genetically distinct individuals. And they are constantly renewing themselves, so that every year most of the actors in the stage of evolution are new, and every hundred years all but a few long lived trees are new!

Arising from all this variation at the DNA level, are the diversity of populations, geographical races, species, and higher categories - genera, families, orders, and classes. Wherever we go, be it a northern taiga or a tropical forest, high up in a mountain, or deep in the sea, we expect and find many kinds of plants, vertebrate animals, insects, and mushrooms, and even if we don't see them, we know that there also exist a large diversity of microorganisms, organized into communities and ecosystems.

Possible Consequences of the Loss of Biodiversity

This enormous diversity is being reduced today by human actions that transform natural landscapes, and transform habitats. These activities include direct actions such as farming, forestry, and livestock raising, and activities as industrialization whose by-products may poison the soil and water, the production of toxic mine tailings, and the deposition on plants, soil, and lakes of chemical radicals through acid rain resulting from industrial air pollution.

The loss of biodiversity resulting from these various human activities which we call Global Change has alarmed many people, and has created a world-wide movement to save the tropical forests, and to conserve as much of the World's Biodiversity as possible (Mann 1991). Some writers (Myers 1979, 1989; Ehrlich and Ehrlich 1981) have insinuated that the loss of biodiversity threatens ecosystem integrity, and indirectly perhaps human existence itself. But although there is a great deal of fear about the possible negative effect of human activities on biodiversity, the fact is that we have very little quantitative evidence regarding the postulated negative consequences of development.

We wish to explore in this introductory chapter whether the arguments that have been brought forth regarding biodiversity and global change are scientifically valid. We would like to see whether there is enough evidence to maintain that a certain level of diversity is essential for the function of the biota and human society. If this tremendous diversity of living forms that surrounds us, is reduced by a tenth, or a third, or even half, will a point come when human life as we know it will no longer be possible? Or is there enough redundancy in the system that it can be argued that the system will continue to function? We also wish to explore how predicted climatic change will impact ecosystems that have lost part of their diversity. There are those that maintain that although human induced changes in the recent past have not fundamentally affected ecosystem function, that the loss of species has affected the ability of plant and animal communities to respond to large global changes such as the predicted change in climate.

When we observe the beautifully manicured fields of Holland, we notice a great deal of biodiversity. Many different crops are grown, many distinct and beautiful trees are cultivated, there is a great deal of diversity in domestic animals, as well as wild flora, birds, fishes, insects, and other invertebrates. However over the centuries as this country was developed by the Dutch people some biodiversity was lost. At some time in the past bears, herds of deer and packs of wolves must have roamed this area when it was covered largely by forests. The coastal areas were scattered with wetlands and must have supported large flocks of birds of many species. The same of course applies to most of the northern Hemisphere. The fertile fields of Pennsylvania, Ohio, Indiana, and Michigan, that today help to feed the United States and produce an overabundance of maize and pork, where once covered by forests of beeches, oaks, maples, and chestnuts, in which there was a great abundance of all manner of animal species. Many subtropical and tropical areas, especially in the Old World, have also been transformed over the centuries by the civilizations of India, Thailand, China, Japan, and others. Also in the Americas the Mayan and Inca civilizations transformed their landscape. Disease checked the growth of those human populations until relatively recently, limiting the area in which they were active, the size of their population and the extent of the transformation they contrived. This is no longer so, and today we are witnessing in the tropics the same kind of wholesale landscape transformation that Europe saw starting in the 16th century, and North America in the 19th and 20th centuries.

These changes began about ten thousand years ago with the invention of agriculture, relatively recently in terms of the history of the human species (Solbrig 1991). They accelerated with the early development of capitalism in the 14th to 16th century that had its origin here in the low countries of what is today Belgium and Netherlands (Braudel 1981), and accelerated still further with the industrial revolution in the last century. The transformation of the landscape was drastic in many areas, and few were affected more than the Netherlands. Marshes and wetlands were drained, large areas of land were gained from the sea, forests were cut, fields were levelled, and every effort made to allow only economically important crops to grow in the fields. In the process many species of plants and animals were undoubtedly lost, although we don't know how many. Most species are insects or fungi, and small invertebrate animals, and we have no way of estimating how many existed in Europe before the introduction of agriculture.

Can we honestly say that the changes that have taken place have negatively affected the functioning of the ecosystem? It does not appear to be so. Of course we cannot answer that question rigorously because we do not know exactly what was here before humans transformed the landscape, nor how it functioned then. But human living conditions seem to have improved, and there are no signs of imminent collapse of this society. Nevertheless the Dutch and people all over the world are worried, both about their own local environment and about worldwide deforestation and tropical landscape transformation. Why? I should quickly add that not everybody shares this concern, and that there are those that wish to push development further and as fast as possible. Many politicians in the United States are of this opinion. At the root of the present argument, lie different perceptions about how the natural world functions and is put together.

Ecosystem Function

Until recently, for the public and the vast majority of scientists themselves, the world was viewed as a harmonious system that was in a state of equilibrium or near it. According to this conception, the diversity of the living world is necessary for its proper functioning. We all have heard about "the balance of nature," an ideal state of nature in which every element is in equilibrium with every other. Yet, this ideal state is not what we observe. The weather is constantly changing. The proportions and even the sorts of plants and animals around us are perpetually fluctuating. Mountains get eroded, and lakes get silted. If we go back some 20 000 years, northern Europe was covered with ice. Yet, this idea of a balance, of an equilibrium in nature, has persisted. It dominates the thinking of much of conservation management, which endeavors to reduce disorder and create undisturbed environments. The supporters of the notion of the balance of nature maintain that ecosystems, although not now in balance, are moving constantly toward equilibrium. Balance is seen as the ideal state. Ecosystems are supposedly prevented from attaining it because of exterior forces, which thus are called disturbances. Storms, pest outbreaks, fires, human induced changes, are examples of disturbances that are presumed to keep ecosystems from reaching equilibrium.

The concept of the balance of nature was strengthened in the 18th century by the Newtonian revolution in physics. Newton introduced very successfully the notion that nature could be interpreted in terms of a few simple laws. This so called "Newtonian paradigm" dominated the thinking of most physicists and scientists until recently. Complexity and disorder was thought to be only apparent. According to the Newtonian view the world is formed by elements that are simple and possess a regular and deterministic dynamics (Nicolis and Prigogine 1989; Solbrig & Nicolis 1991).

Today physicists have decided that nature is not simple in the Newtonian sense, but that it is complex, involved, unpredictable. They also have concluded that the Newtonian view is insufficient as a general explanation of how the Universe functions. Physicists are finding that the entire Universe is not in balance. Disturbances, irregularities of all sorts, are no longer seen as aberrations, but as integral parts of nature. Astrophysicists have discovered that at the very origin of the Cosmos, at the so called "big-bang," a singular, irreversible, and complex Universe was produced (see chapter by Nicolis). Similarly, ecologists have been observing and documenting that most if not all ecosystems are not in balance (DeAngelis & Waterhouse 1987). Fires, storms, even hurricanes are and have been part of life since the beginning. Without them, ecosystems as presently put together cannot function properly.

One of the characteristics of complex systems is surprises. In a complex and fluctuating Universe unpredictable things happen. Seen from this point of view diversity is the result of the very complicated dynamics of living systems. Apparent disorder and randomness at one level of integration can produce order at a higher level, not because of a particular deterministic behavior of each and every element, but as a result of the collective random interactions of all elements that imparts a higher level dynamic order to the system. In biological terms, some of the most interesting studies are those of pest outbreaks, such as the spruce-budworm, investigated by Bus Holling (Ludwig *et al.* 1978), and Robert May (1977). In a fluctuating and complex world we no longer can predict what the result of losses in species, and changes in ecosystems types will be. It is very likely that the loss of part of the tropical forests of the Amazonian basin, will not have catastrophic effects on the integrity of the biosphere, anymore than the loss of forests in much of Asia has had. But then we cannot ever be certain. There very well could be surprises. And of course there will be many distinct local effects, on soil properties, drainage patterns, and possibly also local rainfall patterns. Some of these local effects could be very damaging for human populations that depend on them.

Economic Effects of Biodiversity Losses

Although losses in biodiversity may not affect the stability and overall productivity of ecosystems, they still could have devastating economic effects. Until very recently there was universal consensus regarding the benefits of landscape transformation: more food, more wealth, more economic activity resulting in higher material living standards for a growing human population. This thinking was based on the idea of an ever expanding economy built on an ever expanding resource base. As long as a considerable portion of the globe was thinly settled by humans, such economic model was not only feasible, but successful. It also was very inefficient. Land was wasted and degraded, forests cut and destroyed in order to obtain a few economically valuable species, fisheries exploited without consideration to their replenishment. If one mine, or oil field, or forest, or agricultural field, stopped producing, there always was another one further away. Today we are becoming aware that we are living in a finite world, where resources can become exhausted. How far we are from that state we don't know (see chapter by Colwell) but this issue is worrisome, because of the very small resource base on which we depend.

The number of species of plants and animals on which human society depends is very narrow. Twenty species of plants and five species of animals account for over 90% of all human sustenance and international commerce in foodstuffs (table 1). Three cereal plants wheat, rice, and maize account for 49% of human calory intake. If we increase the list to 100 species, we cover 98% of important economic plants and animals, and if we enlarge the list to a thousand species we include essentially most cultivated and useful plants and animals, excepting some ornamentals as well as species used in folk medicine. A thousand species out of 10 million, that is only 0.01% of all biodiversity has some human economic value. That is a very narrow base. The real challenge is to maintain the integrity of these species in a fluctuating world, made more hostile by human activities. Contrary to wild species, cultivated species are dependent on people, who manipulate them. Even their genetic diversity is controlled by human activities. As we transform landscapes, cut down forests, drain wetlands, and change the composition of the atmosphere, we might be creating a scenario for a global ecological surprise. Global warming, decreases in stratospheric ozone, increases in free radicals, increases in atmospheric CO_2, which individually may not be threatening, collectively could produce unsuspected dynamic changes that could endanger seriously our crops and our domestic animals on which we depend.

Knowing the features of non-equilibrium systems is important for managing crops and ecosystems. Non-equilibrium ecosystems do not necessarily return to their previous state when the convulsion is removed. In many cases it is impossible to return them to their original condition. This means that one set of management guidelines will not suffice for all situations. The previous history, and the present disturbance regime will decide the consequences of any management.

Table 1. The twenty most important species in terms of production and area cultivated (data from FAO)

Crop	Area (1000 Ha)	Production (1000 Tm)
Wheat (*Triticum* spp.)	229 347	505 366
Maize (*Zea mays*)	131 971	488 500
Rice (*Oryza sativa*)	144 962	472 687
Potatoes (*Solanum tuberosum*)	20 066	300 616
Barley (*Hordeum vulgare*)	78 698	176 574
Cassava (*Manihot esculentum*)	14 010	135 551
Sugar cane (*Saccharum officinale*)	23 676	121 524
Sweet potato (*Ipomeas batatas*)	7 880	110 651
Sorgum (*Sorgum* spp.)	91 859	104 592
Soybeans (*Glycine soja*)	52 683	100 809
Grapes (*Vitis vinifera*)	9 564	60 297
Cotton (*Gossypium* spp.)	34 712	49 712
Oats (*Avena sativa*)	25 288	49 630
Coconuts (*Cocus nucifera*)		41 040
Rye (*Secale cereale*)	16 738	32 288
Peanuts (*Arachis hypogea*)	18 728	20 708
Beans (*Phaseolus* spp.)	25 665	14 909
Peas (*Pisum sativum*)	8 832	13 199
Tobacco (*Nicotiana tabacum*)	4 111	6 559
Coffee (*Coffea arabica*)	10 574	6 006

Conclusions

Summarizing this brief introduction, diversity as such is not a resource, but a property of living systems. There is so much biological diversity that its extent is hard to grasp. Even if we manage to reduce the number of species by a third, or even a half, there still will be millions of species of plants, animals, and microorganisms. For ethical and esthetic reasons we might not wish to do that, and I personally feel that every effort should be made to conserve species. But given that there is so much diversity, and that natural systems are highly complex and non-linear, there is probably no imminent danger of creating a change that will destroy life on Earth. However humans depend for the their sustenance not on biodiversity as such, but on a diminutive portion of the world's species, less than 0.01% of them. Drastic environmental changes could threaten all or part of this diversity. That is probably the greatest present danger.

But the overwhelming issue is our ignorance. Ignorance about how many species live on the Earth, ignorance about the relation between habitat transformation and species loss, and above all ignorance about what effect the loss of species will have on ecosystem function. I hope that this symposium and the IUBS-SCOPE-UNESCO program in Biodiversity (Solbrig 1991b, c) will reduce some of that ignorance.

References

Braudel, F. 1981. *Civilization and capitalism*. 3 vols. New York: Harper & Row.

Crow, J. F. and M. Kimura. 1970. *An introduction to population genetics theory*. New York: Harper & Row.

DeAngelis, D. L. and J. C. Waterhouse. 1987. Equilibrium and non-equilibrium concepts in ecological models. *Ecological Monographs* 57: 1-21.

Ehrlich, P. R. and A. H. Ehrlich. 1981. *Extinction: The causes and consequences of the disappearance of species*. New York: Random House.

Ehrlich, P. R. and E. O. Wilson. 1991. Biodiversity studies: science and policy. *Science* 253: 758-762.

Goldschmidt, R. B. 1945. Mimetic polymorphism, a controversial chapter of Darwinism. *Quarterly Review of Biology* 20: 147-64; 205-30.

Lima-de-Faria, A. 1988. *Evolution without selection: form and function by autoevolution*. New York: Elsevier.

Ludwig, D., D. D. Jones and C. S. Holling, 1978. Qualitative analysis of insect outbreak systems: The spruce budworm and forests. *J. Animal Ecol.* 47: 315-332.

Mann, C. C. 1991. Extinction: Are ecologists crying wolf? *Science* 253: 736-738.

May, R. 1977. Thresholds and breaking points in ecosystems with a multiplicity of stable states. *Nature* 269: 471-477.

May, R. 1988. How many species are there on earth? *Science* 241: 1441-1449.

Mayr, E. 1963. *Animal species and evolution*. Cambridge: Harvard University Press.

Myers, N. 1979. *The sinking ark*. Oxford: Pergamon Press.

Myers, N. 1989. *Deforestation rates in tropical forests and their climatic implications*. London: Friends of the Earth.

Nicolis, G. and I. Prigogine. 1989. *Exploring complexity*. New York: Freeman.

Solbrig, O. T. 1991a. Ecosystems and Global Environmental Change. In : R. W. Correll and P. Anderson (Eds.), *The science of global environmental change*, pp. 97-108. Berlin: Springer.

Solbrig, O. T. 1991b. *Biodiversity. Scientific issues and collaborative research proposals*. Mab Digest 9, 77 pp. Paris: Unesco.

Solbrig, O. T. (ed.) 1991c. *From genes to ecosystems: A research agenda for biodiversity*. Cambridge, Mass.: IUBS.

Solbrig, O. T. and G. Nicolis. 1991. *Perspectives on biological complexity*. Paris: IUBS.

4. Dynamical systems, biological complexity and global change

Gregoire Nicolis

Introduction

One of the most challenging questions confronting Science and Society today is the extent to which the future states of the geosphere-biosphere system can be predicted in a reliable manner on time scales of biological, economical and societal relevance.

It has been recognized for some time that the possibility to make such predictions and to use them as a reliable input to Global Change and to Conservation and Land Management research is hindered by a number of formidable difficulties. Still, with the launching of increasingly ambitious international programs a consensus has gradually built based on the idea that these difficulties arise from the fact that the geosphere-biosphere system features a large number of variables and poorly known parameters blurring some fundamental underlying regularities. In other words, one identifies the complexity of the system to the "complication" and temporary drawback arising from incomplete knowledge. In this perspective a question of obvious relevance, is, therefore, to gather as much as possible of this missing information. Once this is available the next natural step would be to develop large numerical models incorporating the maximum amount of variables and parameters allowed by modern computer technology, and study the response of the biosphere when some of these parameters (such on carbon dioxide or chlorofluorocarbon concentration) are varied, particularly as a result of anthropogenic activities. These two goals constitute the core of current research on the response of biological systems to Global Change.

Implicit in the above view is the idea that a system (like the geosphere-biosphere) governed by a given set of laws will follow a well-defined unique history, and that a change in the prevailing conditions (caused, for instance, by man's activities) will elicit a response that is roughly proportional to the strength of the change. In the present Chapter it is shown that this way to envision the geosphere-biosphere system is incomplete. In particular, it overlooks the possibility that both the biota and our physical environment are dynamical systems generating complexity as a result of the very nature of the laws to which they are subjected. In other words, it would seem that the complexity of the geosphere-biosphere system is, to a large extent, an intrinsic property, rather than a manifestation of incomplete knowledge or of the presence of a large number of variables and parameters.

In the next section the concepts of bifurcation and chaos, two fundamental mechanisms leading to complex behavior in nonlinear dynamical systems, are summarized. Evidence for bifurcation and chaos in the biota is discussed thereafter. It is shown that these phenomena entail a limited predictability of some of the system's key variables. The practical repercussion of this fundamental limitation for Global Change and for Conservation and Land Management Research are briefly assessed in the final Section.

Nonlinear dynamical systems : Bifurcation and chaos

The evolution of a natural system in the course of time is conditioned by two major factors: a set of laws governing the individual elements constituting the system and their interactions; and a set of constraints acting from the external world, which are usually manifested via a number of control parameters. Typical examples of the latter in the geosphere-biosphere system are the solar influx, the concentration of atmospheric trace gases, the birth, death or mobility rates of a set of populations in an ecosystem, and so forth.

Let X_i, $i = 1, ..., n$, be the relevant variables. In accordance with the above considerations we write their rate of change in time in the form

$$\frac{dX_i}{dt} = v_i(X_1, ..., X_n; \lambda, \mu, ...) \qquad i = 1, ..., n \qquad (1)$$

where v_i stand for the evolution laws and λ, μ, ... for the control parameters. An illustration of eqs. (1) in the field of biology is provided by the equations of population dynamics governing the competition between species sharing common resources, or the Lotka-Volterra equations governing predator-prey interactions.

Whatever the detailed interpretation of eqs. (1) might be, a common feature shared by large classes of systems is that v_i are non-linear functions of the state variables. This ubiquity of nonlinearity in nature stems primarily from the numerous feedbacks exerted between the different components of the system, as a result of which the rates of the fundamental processes contributing to the dynamics are no longer constant, but become state-dependent.

Nonlinearity is a source of intrinsically generated complex behavior and unpredictability, in the sense that more that one outcomes of the evolution become now possible (Nicolis and Prigogine 1989). Figure 1 depicts a typical scenario of the way the solutions X of a nonlinear dynamical system behave when a parameter λ built in it is varied. At the values λ_1, λ_2, ..., of the parameter, usually referred to as bifurcation points, the previously prevailing solution becomes unstable to small perturbations and new branches of solution are generated. In general these bifurcation cascades produce a multiplicity of simultaneously available states, to which we refer as attractors (branches (a) to (d) for the value $\lambda = \bar{\lambda}$ in Fig. 1). Which of these states will actually be chosen depends on the initial conditions. This property confers to the system a high sensitivity to parameters and a markedly random character, since the initial conditions are history-dependent and may be modified by the fluctuations or by slight external perturbations. In actual fact therefore the dynamics of a multistable system will be an aperiodic succession of intermittent jumps between coexisting attractors. This view is reminiscent of a great number of natural processes involving abrupt transitions.

A very convenient representation of the states that can be reached by a system beyond a bifurcation is provided by the phase space, the space spanned by the full set of variables X_1, ..., X_n participating in the dynamics (Nicolis and Prigogine 1989). The nature of the phase space "portrait" depends on whether the system is conservative or dissipative. Experiment shows that the great majority of systems encountered in nature are dissipative, a property that shows up through an irreversible evolution to a preferred set in phase space, which we call attractor. Attractors enjoy the important property of asymptotic stability, that is to say, the ability to damp perturbations. This in turn ensures a certain degree of reproducibility of the behavior.

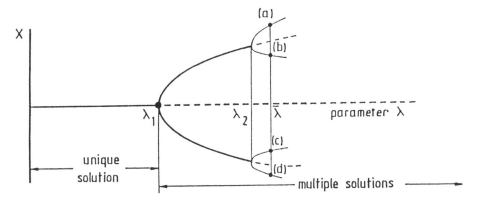

Figure 1. Typical bifurcation diagram of a nonlinear dynamical system.

Whatever its detailed nature, an attractor has a measure in phase space which is equal to zero. In other words, in a phase space of n dimensions its dimensionality d will satisfy the strict inequality $d < n$. Actually, in a great number of cases d is much less than n, indicating that a drastic reduction of the description of the system is possible. The simplest attractors are zero dimensional (point) and one dimensional (limit cycle) manifolds, as depicted in figure 2. However, one now realizes that in many instances bifurcation cascades lead the system to the regime of deterministic chaos alluded already in the Introduction (Baker and Gollub 1990; Bergé et al. 1984). Figure 3 depicts a typical attractor describing chaotic dynamics, obtained by numerical integration of a three-variable model system (Rössler 1979). Figures 4a and 4b illustrate two ubiquitous features shared by all chaotic attractors: first, contrary to what happens in a periodic, limit cycle attractor (Fig. 2b) the trajectory never closes back to itself (Fig. 4a); and second, in addition to certain well-defined characteristic frequencies a chaotic attractor generates a broad band spectrum reminiscent of random noise (Fig. 4b). Chaotic attractors thus show that irregular, aperiodic behavior sharing some features of a random process may be generated by a dynamical system governed by perfectly deterministic laws of evolution.

Let us have a closer look at the structure of the attractor in figure 3. We observe two opposing trends. On the one hand (see horizontal arrow) an instability of motion tending to remove the phase space trajectory away from the "reference state" $x = y = z = 0$; and on the other hand (see vertical arrow in the figure) the bending of the outgoing trajectories followed by their reinjection back to the vicinity of this state.

The unstable motion on the attractor (as opposed to the stability in the directions transverse to the attractor) is reflected by the sensitivity of the trajectories on the attractor to minute changes in the initial conditions, as a result of which two initially nearby states diverge, on the average, in an exponential fashion. This confers to the evolution an irreducible element of unpredictability. The characteristic rate of divergence σ_L is referred to as the (largest) positive Lyapounov exponent of the system. For the model of figure 3 it turns out that $\sigma_L = 0.15$ bits per unit time.

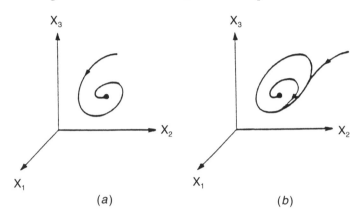

Figure 2. Evolution toward a point attractor (a) or limit cycle attractor (b) in a dissipative dynamical system.

In general the Lyapounov exponents σ_L cannot be computed analytically. Algorithms are currently available, allowing their determination from the knowledge of a time series pertaining to the evolution of a single variable. The reader is referred to the monograph of Baker and Gollub (1990) for a recent survey of this topic.

Bifurcations and chaos in the biota

Traditionally, the link between geosphere and biosphere is seen as being of the "cause-effect" type whereby the latter is first driven passively by the former and eventually feeds back on it essentially through a change of parameters. There is, however, ample evidence of a far deeper interdependence. As pointed out by Lovelock (1987) the current composition of the air and oceans is a highly nonequilibrium one. Since the composition of lifeless planets like Venus and Mars seems to be quite close to the equilibrium one, it is legitimate to stipulate that life has literally shaped our Global Environment and helped since then to maintain it, together with the incoming solar influx, to a highly nonequilibrium state.

A nonequilibrium environment is apt to evolution and change, since the rate of a process (like, for instance, an innovation) need not be counteracted by the rate of the reverse process (as it happens in equilibrium). The conditions for the phenomena of instability, bifurcation and chaos summarized in the preceding section are, therefore, met.

It has been recognized for some time that bifurcation is at the origin of such important biological phenomena as rhythmic behavior at both the subcellular and supercellular level, or the emergence of spatial patterns of differentiated cells in an initially homogeneous morphogenetic field. Closer to the preoccupations of the present volume, as genetic information is redundant there are many more genotypes than phenotypes, leaving room for selectively neutral variants. Environmental factors and transition phenomena become thus relevant in this case, as it is impressively demonstrated by the fact that many primitive organisms, most bacteria for example, unfold different reproducible phenotypes in different environments. The idea can be pushed even further. As pointed out by Eigen and Schuster (1979, and Schuster 1991)

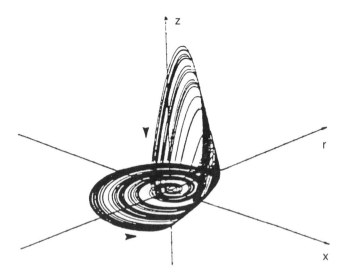

Figure 3. Chaotic attractor obtained from numerical integration of the system of eqs. for parameter values a = 0.38, b = 0.3, c = 4.5.

the spatial structure (secondary, 2-d or tertiary, 3-d), of a nucleic acid molecule like RNA can be considered as its phenotype, whereas the sequence of nucleotides along the chain stands for the genotype. The driving force for the formation of a phenotype from the genotype is then reduced to a universal thermodynamic principle of free energy minimum governing the folding of the molecule and entailing that the dynamics in phase space can be viewed as a "hill climbing" on a landscape provided by an optimizing functional (Fig.5). In the real world such "fitness landscapes" are highly complex and are for this reason usually referred to as "rugged value landscapes" . They are characterized by the existence of many local extrema covering a broad spectrum of heights. Evolutionary optimization at the molecular level at the very basis of biodiversity at the macroscopic level can thus be visualized as a search for low lying minima preferably, the global minimum. The typical mechanism of this search is bifurcation. Whether the global optimum will actually be reached is a matter of the structure of the specific fitness landscape, the time span available and the population size. What is certain is that the sequence of states visited by the system in the course of such an optimization process will be markedly unpredictable as it will depend on minute differences in the values of environmental factors (such as pH) and on the intrinsic variability associated to mutation

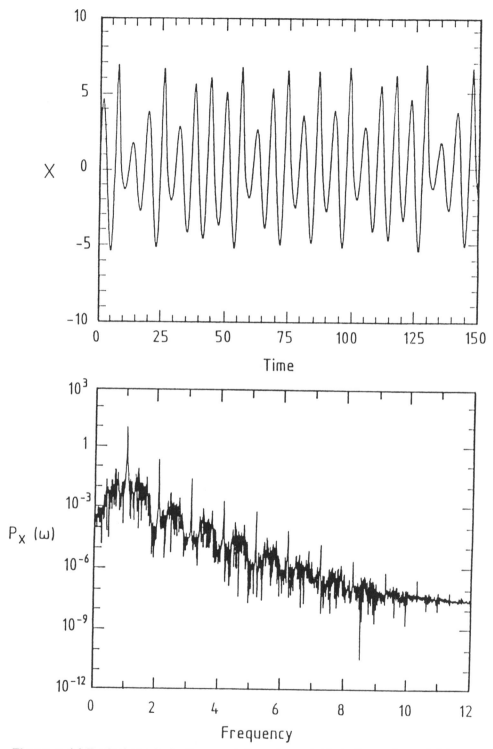

Figure 4. (a) Typical aperiodic time evolution of a variable in the regime of deterministic chaos; (b) power spectrum associated to this variable, displaying the broad band character usually attributed to random noise.

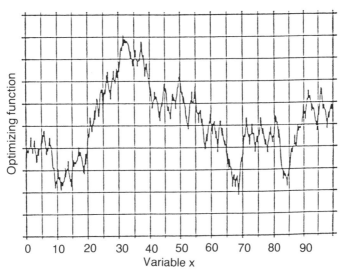

Figure 5. A typical rugged value landscape.

rate. It is, therefore, hardly an exaggeration to state that bifurcation is a ubiquitous mechanism at the very basis of biodiversity.

Nonlinear behavior and bifurcations may also be prominent at the macroscopic level of population dynamics in an ecosystem. Multiple steady states in plausible models of predation have been suggested by Holling (1965, 1973). This possibility is of obvious relevance in resource management policies. More surprisingly perhaps, recent inves-

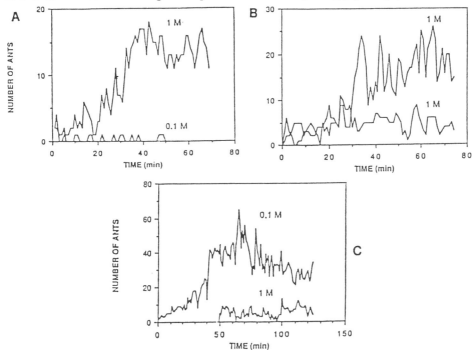

Figure 6. Response of a colony of L. niger to the discovery of food sources. (A) 1M and 0.1 M sucrose sources introduced simultaneously; (B) Two 1M sources introduced simultaneously; (C), 1M source introduced 50 mins after a 0.1 M source.

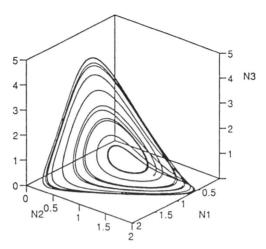

Figure 7. Chaotic attractor associated to Eqs. (2). Parameter values : $K = 1$, $a_{11} = a_{12} = 0.5$, $a_{13} = a_{23} = a_{32} = a_{33} = 0.1$, $a_{21} = -0.5$, $a_{22} = -0.1$, $a_{31} = 1.45$.

tigations suggest that the capacity of an animal group or of a society for organizing its activity and adapting to its environment can spring from the autocatalytic interactions between individuals and the interactions between the individuals and the environment. A striking illustration is provided by foraging in social insects.

When new food sources are discovered simultaneously by a group of ants, recruitment of foragers is started. Responding to the food sources' different quality the foragers lay more or less pheromone when returning to the nest. The recruitments proceed at different rates, and this process generates competition for inactive foragers waiting in the nest to be recruited. This competition can generate complex social decisions well beyond the capacity of an individual.

Experiments combined with modelling have shown that trail laying ants can use their trail recruitment to choose the richest food source (Pasteels and Deneubourg 1987). For example, *Lasius niger* foragers when offered simultaneously a 0.1M and a 1M sucrose solution concentrate their activity on the 1M source (Fig. 6A). When offered two identical 1M sources simultaneously they concentrate on one of them rather than exploiting both equally (Fig. 6B). The particular source to be selected depends entirely on small random deviations from a completely symmetrical distribution of the populations initially visiting the food sources, which are inevitable in a real world system. This is completely equivalent to the phenomenon of bifurcation. Notice that individuals can become prisoners of their trail system in the sense that once a trail is well-established, a new trail is unable to compete with it and develop, even if it leads to a richer source. When offered just a 0.1M source *L. niger* foragers establish a trail to it and exploit it. If you then add a richer 1M source, they discover it but are incapable of switching their activity to it (Fig. 6C). This illustrates quite clearly that the search for optimality is not always the most appropriate model for understanding collective, autocatalytic behavior at the macroscopic level.

Population biology provides also compelling examples of chaotic behavior. Mathematical models of population growth and competition produce chaotic behavior most naturally, owing to the fact that the equations for the rate of change of a population at a given time depend on the state at previous generations, which is separated from the present state by a finite time lag (May and Oster 1976). Experimental data are not as clear-cut, mainly because of the large environmental noise that is inevitably superimposed on the dynamical behavior. Still, they do suggest that irregular variations in time and space are ubiquitous (Schaffer and Kot 1985).

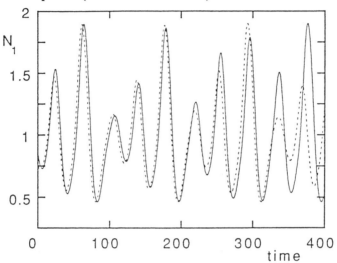

Figure 8. Illustration of the concept of unpredictability beyond the Lyapounov time.

Let us analyze the consequences of chaotic behavior on a simple prototype model. We consider three coupled Volterra-Verhulst type equations,

$$\frac{dN_i}{d_f} = N_i \sum_{j=1}^{3} a_{ij}(K - N_j) \qquad i = 1,2,3 \qquad (2)$$

where N_i denote the (normalized) population densities of three species sharing common resources, a_{ii} their growth rates, and a_{ij}, $i = j$, the effect of interactions arising from niche overlap. It can easily be established that in certain ranges of values of the parameters $\{a_{ij}\}$ eqs (2) give rise to chaotic behavior. The corresponding attractor is depicted in figure 7.

Figure 8 illustrates on this model the concept of limited predictability introduced in the previous section. The system is started with two slightly different initial conditions. This deliberately introduced difference is supposed to represent the finite precision that limits any measurement process. To the observer, therefore, the two initial states are indistinguishable. As shown in figure 8 they remain so for some time, but eventually (here after about 200 time units) they begin to split until their difference reaches a macroscopic level. Confronted with this situation the observer will conclude that the system at hand is unpredictable beyond a certain lapse of time. The latter is nothing but the inverse of the positive Lyapounov exponent, which is an intrinsic property of the dynamics (eqs (2)) once the parameter values are specified. We thus see how systems governed by completely deterministic laws can behave in a manner that is practically indistinguishable from a stochastic process.

How does a system showing complex behavior respond to environmental change? Is its response passive, cause-to-effect like as stipulated in typical Global Change studies or is the situation more subtle? Figure 9 provides a tentative answer to this question. We take one of the parameters (here a_{31}) to be subjected to a slow time increase (much slower than the system's intrinsic time scales) reflecting a gradual drift in environmental conditions and study, using again eqs. (2), the time dependence of the

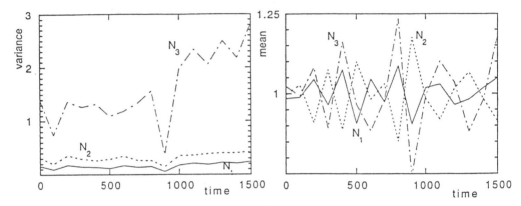

Figure 9. Response of the system of eqs. (2) to a slow variation of the parameter a_{33}, $a_{33} = 1.45 + 0.0001t$. Upper and lower panels denote, respectively, the time behavior of mean and variance smoothed over 100 time units.

variables under these conditions. The right panel of figure 9 depicts the time dependence of the "smoothed" variables, averaged over 100 time units. The system is again seen to perform random looking excursions in both directions around its long term average (despite the systematic increase of a_{31}). As a matter of fact one can hardly detect any drift whatsoever on the envelope of these complex oscillations.

Let us now study the response of the variance (averaged, again, over 100 time units) around the above mean. As seen in the left panel of figure 9 N_1 and N_2 are hardly affected but the effect on N_3 is quite dramatic indeed. Although it features an important systematic drift upward it is still subjected to variability, as a result of which during substantial periods of time the upward trend is actually reversed. This shows how cautious one has to be in predictions about the effect of parameters on a complex system. It also suggests that since the variance is a more sensitive probe of change than the mean it should be seriously considered as a quantity to be monitored in future experimental studies.

Concluding remarks

In this chapter the possibility of spontaneous transitions and of complex behavior in the geosphere-biosphere system arising from the nonequilibrium conditions and the nonlinear character of the underlying laws of evolution has been suggested.

One of the major conclusions of our analysis is the existence of intrinsic limits in monitoring and predicting future states of a system undergoing bifurcations and chaotic dynamics. This imposes severe constraints on the usefulness of large numerical models currently used in the study of anthropogenic effects in the biota. Indeed, in view of the sensitivity on parameters and/or on initial conditions, the particular time and space history predicted in a given run of such a model may well turn out to correspond to a completely fictitious world that has nothing to do with the actual evolution of our own geosphere-biosphere system during the same lapse of time. Curiously, the statistical properties of the system display some striking regularities that turn out to be much more robust than the local properties.

The ideas put forth in this Chapter should also have serious repercussions in the observational strategies that are being developed in the framework of various international programs. Great emphasis is currently placed on data collection and management. Data are of paramount importance, but one should be aware that whereas in some sectors data are scarce, in some other sectors there are overabundant. Further-more, data collection must be placed in a certain perspective. The number and type of data to be collected in order to obtain, say, a power spectrum are different from those needed, say, to reconstruct the dynamics in a finer manner. In a similar vein, the spatial distribution of stations of an observational network becomes a very important issue in view of the highly irregular and intermittent spatio-temporal variation of atmospheric, oceanic and ecological fields. All in all it is our conviction that a drastic reassessment of attitude is needed from which a more realistic, more constructive and undoubtedly more challenging view of our biological and physical environment is bound to emerge.

Acknowledgments

I am indebted to J.L. Deneubourg and R. Beckers for interesting discussions and to the European Commission (EPOC-CT90-0012 (TSTS) project) and the Belgian Government (GLOBAL CHANGE program) for financial support.

References

Baker G. and J. Gollub, 1990. *Chaotic dynamics*. Cambridge: Cambridge Univ. Press.

P. Bergé, Y. Pomeau and C. Vidal, 1984. *L'Ordre dans le chaos*. Paris: Hermann.

Eigen M. and P. Schuster, 1979. *The Hypercycle*. Berlin: Springer.

Holling C.S. 1965. The functional response of predators to prey density and its role in mimicry and population regulation. *Mem. Ent. Soc. Canada* 45: 1-60.

Holling C.S. 1973. Resilience and stability of ecological systems. *Ann. Rev. Ecol. Syst.* 4: 1-23.

Lovelock, J. 1987. *Gaia*. Oxford: Oxford Univ. Press.

May R. and G. Oster, 1976. Bifurcation and dynamic complexity in simple ecological models. *Am. Nat.* 110: 573-599.

Nicolis G. and I. Prigogine, 1989. *Exploring Complexity*. New York: Freeman,

Pasteels J. and J.L. Deneubourg (eds.), 1987. *From individual to collective behavior in social insects*. Basel: Birkhèuser.

Rössler, O. 1979. Continuous chaos - four prototype equations. *Ann. New York Acad. Sci.* 316, 376.

Schaffer W. and M. Kot, 1985. Do strange attractors govern ecological systems? *Bioscience* 35: 342-350.

Schuster, P. 1991. Optimization dynamics on value landscapes. In O. T. Solbrig and G. Nicolis (ed.), *Perspectives in biological complexity*, pp. 115-161. Paris: IUBS.

5. Biodiversity at the molecular level

Jean-Claude Mounolou

We tend to believe that humans have become a keystone species of the biosphere. In doing so they have progressively taken over in their hands the role of thousands of species that have modelled ecosystems through a continuous change since life appeared on earth. One obvious consequence is the rapid reduction of biodiversity (deforestation being a critical example). In order to evaluate and eventually master this drastic event (although recent in terms of evolutionary times) an assessment of the role of biodiversity in ecosystem function is required. When reviewing this question F. Di Castri and T. Younès (1990) concluded that little is known and developed a series of problems that challenged the scientific community about the origins of biodiversity, the degree of species redundancy, the thresholds of irreversibility and the extinction processes, and the general occurrence of keystone species.

Three contributions in this field can be expected from molecular biologists. The first and most extensive one provides a thorough identification of biodiversity at the individual level (organism, cell, genotype and phenotype) and of its activity and adaptive potential (physiology, regulations). The second yields information about molecular processes that generate, maintain and reorganize them. The last contribution is more speculative since very few if any experiments have been carried on in this regard yet. Under the spur of ecologists molecular biologists are bound to reorientate their comfortable reductionist approach to such questions as what molecular parameters enable a species to become a keystone one and how will it evolve in consequence of its own critical position.

Biodiversity among organisms stems from variation in the sequence of the DNA they carry and from their adaptive response to the environment. DNA is unique as its molecular structure provides genetic information and its metabolism insures replication and mutation. The former is the basis of biological stability and individual expansion through faithful multiplication. The latter generates diversity. As genetic information is organized in numerous specific genes, the shuffling of genes, through the molecular recombination of the DNA at the time of sexual reproduction or of horizontal genetic transfers carried on by viruses or interspecific crosses, generates diversity at the individual level more rapidly than the mere accumulation of mutations would do. Placed in a varying environment cells and organisms adapt their activity within the limits allowed by their genotype. Several cellular mechanisms insure both short term and long term responses to external stimuli. The first ones are best analyzed in terms of flow of metabolites and metabolic controls (Kacser and Burns 1973) or of enzyme activities (activations, inhibitions, chemical modifications). The long term adaptation is provided through the complex mechanisms of the regulation of gene expression (induction, repression, post-transcriptional controls) and of cellular compartimentation which ultimately can provide examples of self-perpetuated regimes through membrane elaboration and activities. In the classical view expanding diversity reduces the probability of genetic exchange between some individuals and ultimately leads to the emergence of new species. Random drift and selection acting on the level of organisms and communities constantly reorganize the extent of biodiversity both geographically and numerically. They provide a changing biological landscape that has integrated the potential of previous and present genotypes, the selective effect of their environments as well as random founding effects and encounters. Biochemical tools (PCR, sequencing techniques, antibodies) are now available to characterize diversity, monitor the changes and establish the molecular basis of population biology (illustration in Avise *et al.* 1987) and the phylogeny of living organisms (Hori, chapter 10).

This comfortable and satisfactory picture well presented in modern molecular and cellular biology text-books may be misleading for two reasons. On one hand its synthetic presentation hides unknown aspects of biological complexity at the molecular level and on the other it does not provide efficient and specific experimental proposals to tackle the fundamental questions on the role of biodiversity in the function of ecosystems. Three recent developments in molecular biology may illustrate this point and force new speculations.

Genomes

What makes a genome coherent and the cell that harbors it consistent is a fundamental question molecular geneticists are now in a position to try to answer. Table 1 shows that genomes that enable cells (or organisms) to be self sufficient vary considerably in size from one species to the other. At first genome complexity seems to parallel that of the organism's organization and its position in evolutionary phylogenies. However remarkable exceptions are noticeable (fungi are an example) and raise the question of the exact amount of information needed to sustain various forms of life. Furthermore molecular biologists know that most of size variation is due to the accumulation of repeated and apparently useless sequences (selfish DNA) in some genomes. In this respect, a great deal of information is expected to be produced by the national and international programs that envision for the turn of the century the formidable task of deciphering the complete sequence of several genomes. The Human Genome project is the ultimate one (coordinated by HUGO, Roberts 1990; Maddox 1991) but other genomes are also under intensive study (e.g. yeast, *Bacillus subtilis, Caenorhabditis elegans, Arabidopsis thaliana*). At present what has emerged from these programs is already extremely interesting and will force us to reconsider the molecular approach to biodiversity. The immediate benefit is the implementation of data bases providing reference sequences of many genes. Among those some alleles which are responsible for human hereditary pathologies are the most actively investigated and IUBS supports the exploration of genetic diseases among humans through the Decade of the Tropics Program (Roberts, 1990). Those sequences also serve as a molecular base to describe genetic diversity in general and to evaluate its extent. They provide a reservoir for microorganisms, plants and animals breeding projects using classical genetics and/or recombinant DNA and transgenosis techniques. Their knowledge will constitute a basis for decisions to be implemented in Genetic Resources Conservation Programs.

Yet more is to be expected from the knowledge of genomes sequences as a way of understanding biodiversity. For a long time molecular biologists have speculated about the number of genes that are necessary to build a self-sustaining cell, and about the number of those present in highly evolved plants and animals. Figures usually range from some 10^3 to 10^5. In other words the question is: do we know all the types of genes and proteins acting in a cell or are there still some major discoveries ahead of us ? The recent complete sequencing of yeast chromosome III by a federation of European laboratories directed by EEC opens a decisive path in the field (Oliver *et al.* 1992). Out of several hundred genes on this chromosome 60 % turned out to be completely elusive. They were not known through mutants and classical genetics, their sequences when compared to well-defined gene sequences stored in data banks do not correspond to any group of classical genes (such as genes that code for membrane proteins, kinases, histones, or polymerases). However these genes are *bona fide* ones. They are transcribed in the yeast cell. Disruption of these genes is usually not lethal in laboratory growing conditions. It has been suggested that these genes are probably involved in the modulation of cell activities. They probably constitute the genetic adaptive potential of the cell to respond to environmental changes. The nature of their products as well as their activities are now to be elucidated. But for the biologists this discovery brings a major stone to bridge the gap between cell biology and biodiversity.

Table 1. Genome sizes and Sequencing projects.

Organism	Genome Size	Sequencing Project
Eucaryotes		
Human	3×10^9 bp	HUGO
Mouse	3×10^9 bp	-
Drosophila	1.5×10^8 bp	multinational
Arabidopsis	10^8 bp	US and UK laboratories mainly
Neurospora	4×10^7 bp	-
Nematode	8×10^7 bp	UK and US laboratories
Yeast	1.5×10^7 bp	Mainly European laboratories (EEC), chromosome III is already sequenced
Procaryotes		
Escherichia coli	4.7×10^6 bp	US and Japan
Bacillus subtilis	4×10^6	US and Europe
Chlamydia trachomatis	1.4×10^6	
Mycoplasma	10^6	Europe and US

The second main subject where a contribution of genome sequencing programs is expected deals with repeated DNA sequences. Briefly the situation we face appears to be the result of a basic property of DNA metabolism and of different evolutive strategies used by different species to exploit it. The molecular apparatus that is responsible for DNA replication, recombination and repair allows to repeat and transpose sequences in the genome at relatively low frequencies. In this respect DNA perturbation through replication is not as faithful as it looks at first and genomes are constantly susceptible to changes. This property of DNA metabolism tends to constantly increase the size of the genome and its informational content. But mutations as well as transposition events modify and inactivate these DNA sequences and the genome is progressively loaded with selfish DNA. Very clearly species have exploited differently this basic potential. Microorganisms and fungi have minimized its effects and genomes are relatively short and free of repeated sequences. On the contrary higher plants and animals have used it to generate families of genes as a way to organize different tissues through some divergence and differential regulation of the various members of each family. (A price was paid however and this was accompanied by considerable increase in the selfish DNA load and the genome size). In this field again much is to be excepted for the understanding of biodiversity from some knowledge of the mechanisms underlying these different strategies.

The following examples have been chosen to illustrate various situations where a molecular examination can help increase our understanding of biodiversity and to evaluate its future.

Insecticides resistance of mosquitoes has spread very rapidly after massive treatments of insect populations with these chemicals. In very elegant studies that carry on converging approaches in molecular biology, population genetics, and biogeography N. Pasteur and her group have shown that resistance is the consequence of the repetition in large numbers of an esterase gene the protein product of which inactivates the pesticides (Raymond and Pasteur 1989). The resistant genome spreads out in the populations owing to its advantage, but is not necessarily maintained when the selective pressure is withdrawn (Raymond et al. 1991). Other examples of such a situation are known and it seems that this sequence repetition strategy has also been used to accommodate both external and internal (genotypic) constrains. In the case of ribosomal RNA genes the number of repeats can be rather tightly controlled and their sequences showing concerted evolution are homogenized through a molecular drive process (Dover 1982; Dover et al. 1982). In the case of the noncoding region of rabbit mitochondrial DNA the number of repeated sequences is not tightly defined but molecular constrains due to replication and/or transcription preclude both a decline to unicity and an unlimited increase (Mignotte 1990; Biju-Duval et al. 1991). This leads to the more general question on how a cells counts the number of repeated sequences (including genes) it harbors and how it deals with them.

Traditionally we are taught that in eucaryotes counting is achieved at the time of meiosis through the pairing of homologous chromosomes. Absence of pairing at mitosis has led to the feeling that this event was restricted to and characteristic of meiosis. Early hints that it could be of more general occurrence were provided by the observation of somatic crossing-over by geneticists (Stern 1936). However this was considered as a rare escape to the law. Recent studies are now forcing in the idea that pairing is a more systematic and widespread process although its molecular basis is still not well understood. At the same time they suggest a way to analyze how species have used various strategies to exploit DNA metabolism properties, to create genetic diversity and build up the adaptive potential to respond to environmental changes. In bacteria elegant experiments were devised to bring together in the same cell *Escherichia coli* and *Salmonella typhimurium* genomes (Rayssiguier *et al.* 1989). The expectation was that these two rather homologous and recently diverged sequences would lead to abundant recombinants. On the contrary genetic exchanges were very limited. Using mutations that impair the DNA repair apparatus of the cell the authors were able to show that indeed pairing between the two genomes occurs but that repair of mismatched DNA strands corrects back DNA molecules to parental information even before molecular recombination happens. Consequently the genetic exchange between the two species is considerably limited. In this case the properties of DNA metabolism are clearly exploited in a strategy of specifying and increasing diversity.

In fungi Selker and his colleagues using *Neurospora* (1990) and Rossignol and coworkers using *Ascobolus* (Faugeron *et al.* 1990) have recently shown using recombinant DNA and transgenosis techniques that pairing of repeated homologous sequences is not restricted to meiosis. When it occurs in premeiotic cells (after fertilization) it provides the substrate for a set of molecular events that involve DNA methylation. In the case of *Neurospora* they lead to very efficient mutagenesis of the repeated sequences, inactivation of the corresponding genes and ultimately to lethality. *Neurospora* seems to have devised a strategy of repetitions and consequently of selfish DNA accumulation exclusion. In *Ascobolus* the consequence of the methylation of paired repeated sequences is not so drastic but genes carried on these chromosomal regions are no longer expressed. This silencing is however reversible. Reexpression is under both developmental and environmental control and *Ascobolus* stores archived DNA. In higher plants (e.g. *Petunia, Arabidopsis*) several similar situations, called "cosuppressions" or trans interactions between homologous genes, have been discovered (Jorgensen 1991). Here too the duplication of genes leads to their coordinated inactivation presumably due to molecular processes acting after their paring. The biochemical reactions involved are not yet elucidated and why ribosomal RNA genes or transposable elements escape these processes is not understood.

In principle gene inactivation can also be carried on by trans-acting diffusible factors (mainly proteins) without any pairing. Molecular biologists have provided ample evidence for this and on the reversibility of the event. Methylation of DNA is often involved as well as reorganization of chromatin structure. This has long been documented in the case of heterochromatin X inactivation in mammals discovered by M. Lyon (1969) certainly the most thoroughly analyzed example (Davies 1991). We now know that the process is initiated at a well located inactivation center on the chromosome, how its develops and why some genes are susceptible to it and others are not is still elusive (Brown *et al.* 1991). Similarly when Schaeffer *et al.* (1976) discovered that *Bacillus subtilis* could be (and stay) genetically diploid, provided that one chromosome was silenced, they opened inevitably the path to what is now known as archived DNA (Thaler 1990). In these cells still unknown molecular modifications of the DNA and/or the nucleoid DNA-protein complex prevent the transcription but not the duplication of the chromosome. The interesting point however is that the process does not need a full-length chromosome and that inactivation is transient. Reexpression is induced by environmental changes (such as those leading to sporulation).

All this molecular and genetic information is of considerable interest for biologists. It reveals that the molecular panel of diversity is much wider and more complex than it was thought at first. Moreover when diversity occurs and is selected, the genetic and molecular strategy that is in part responsible for its appearance is also selected (Cairns *et al.* 1988; Hall 1990). In consequence with time biodiversity is still expandable and able to respond to environmental changes even when it has suffered considerable restrictions.

In conclusion as molecular genetics bring in data and prepares concepts, bridging stones between the knowledge of the diversity of cell function and that of biodiversity in the exosystems appear progressively. If the way is not yet paved all along three developments should help it in the near future. One is to draw more interest in a molecular and comparative approach on some critical questions about biodiversity, such as redundancy, thresholds of irreversibility, keystone species molecular characterization, function and evolution. The second is the quantitative evaluation of the adaptive value of molecular mechanisms acting at the DNA level. At last we are now in a position to analyze how genetic diversity expands in time after severe reductions.

References

Avise, J.C., J. Arnold, R.M. Ball, E. Bermingham, T. Lamb, J.E. Neigel, G.A. Reeb, and N. C. Saunders, 1987. Intraspecific phylogeography: the mitochondrial bridge between population genetics and systematics. *Ann. Rev. Ecol. Syst.* 18: 489-522.

Biju-Duval, C., H. Ennafaa, N. Dennebouy, M. Monnerot, F. Mignotte, R. Soriguer, A. El Gaaied, A. El Hili, and J.C. Mounolou, 1991. Mitochondrial DNA evolution in lagomorphs: Origin of systematic heteroplasmy and organization of diversity in European rabbits. *J. Mol. Evol.* 33: 92-102.

Brown, C.J., A. Ballabio, J.L. Ruperet, R.G. Lafreniere, M. Grompe, R. Tonlorenzi, and H.F. Willard, 1991. A gene from the region of the human X inactivation entre is expressed exclusively from the inactive X chromosome. *Nature* 349: 38-44.

Cairns, J., J. Overbaugh, and S. Miller, 1988. The origin of mutants. *Nature* 355: 142-145.

Davies, K. 1991. The essence of inactivity. *Nature* 349: 15-16.

Di Castri, F. and T. Younès, 1990. Ecosystem function of biological diversity. *Biology International* 22: 1-18.

Dover, G.A. 1982. Molecular drive: a cohesive mode of species evolution. *Nature* 299: 111-117.

Dover, G., S. Brown, E. Coen, J. Dallas, T. Strachan, and M. Trick, 1982. The dynamics of genome evolution and species differentiation. In : G. Dover and R. Flavell, (eds.) *Genome evolution*, pp. 343-355. London: Academic Press.

Faugeron, G., L. Rhounim, and J. L. Rossignol, 1990. How does the call count the number of ectopic copies of a gene in the premeiotic inactivation process acting in *Ascobolus immersus*? *Genetics* 124:585-591.

Hall, B.G. 1990. Spontaneous point mutations that occur more often when advantageous than when neutral. *Genetics*. 126: 5-16.

Jorgensen, R. 1991. Beyond antisense - how do transgenes interact with homologous plant genes. *Trends in Biotechnology* 9: 266-267.

Kacser, H. and J.A. Burns, 1973 The control of flux. In D.D. Davies (ed.), *Rate control of biological processes*, pp. 65-104. London: Cambridge University Press.

Lyon M.F., 1969. Gene action in the X-chromosome of the mouse (*Mus musculus*). *Nature* 190: 372-373.

Maddox, J. 1991 The case for the human genome. *Nature* 352:11-14.

Mignotte, F., M. Gueride, A-M. Champagne, and J.C. Mounolou, 1990. Direct repeats in the noncoding region of rabbit mitochondrial DNA: Involvement in the generation of intra and inter individual heterogeneity. *Eur. J. Biochem*. 194: 561-571.

Oliver, S.G. *et al.* 1992. The complete DNA sequence of yeast chromosome III. *Nature* 357: 38-46.

Raymond, M. and N. Pasteur, 1989. The amplification of B1 esterase gene in the mosquito *Culex pipiens* is present in gametes. *Nucleic Acid Res*. 17: 7116.

Raymond, M., A. Callaghan, P. Fort, and N. Pasteur, 1991. Worldwide migration of amplified insecticide resistance genes in mosquitoes. *Nature* 350: 151-153.

Rayssiguier C., D.S. Thaler, and M. Radman, 1989. The barrier to recombination between *Escherichia coli* and *Salmonella typhimurium* is disrupted in mismatch-repair mutants. *Nature* 342:396-401.

Roberts, D.F. 1990 The human genome initiative and the IUBS. *Biology International* 21:3-11.

Schaeffer, P., B. Cami, and R.D. Hotchkiss, 1976. Fusion of bacterial protoplasts. *Proc. Natl. Acad. Sci*. USA. 73: 2151-2155.

Selker, E.U. 1990. Premeiotic instability of repeated sequences in *Neurospora crassa*. *Ann. Rev. of Genetics* 24: 579-613.

Stern C., 1936. Somatic crossing over and segregation in *Drosophila melanogaster*. *Genetics* 21: 625-730.

Thaler, D.S., J.R. Roth, and L. Horschbein, 1990. Imprinting as a mechanism for the control of gene expression. In K. Drlica, and M. Riley, (eds.), *The bacterial chromosome*, pp. 445-456. Washington D.C.: American Society for Microbiology.

6. Genetic diversity and its role in the survival of species

Wilke van Delden

Genetic variation

Biological diversity is sometimes identified solely with number of species, but the concept can also be applied to the intraspecies level. Although species are characterized by having a set of particular common traits, the individual members of a species may vary considerably. Part of this variation will be environmentally induced, but part of it is of genetic origin. The presence of genetic variation is a basic characteristic of species. Phenotypic expression of genetic variation may vary between classical polymorphisms with a limited number of distinct phenotypes, and quantitative characters with a continuous distribution. From an evolutionary viewpoint genetic variation is of fundamental importance for the permanence of a species as it provides the necessary short term adaptation to the prevailing abiotic and biotic environmental conditions, while on the long term it enables changes in the genetic composition to cope with changes in the environment.

A first requisite for the study of the role of intraspecific genetic variation in the survival and extinction of species is knowledge about the level of genetic variation. In genetic terms this means determination of the number of polymorphic loci, the number of alleles, dominance relations, genetic architecture and spatial distribution of genetic variants. Genetic variation at the level of the individual - the number of genotypes - is potentially tremendous. With x varying loci, each with a alleles, the number of genotypes, g, is:

$$g = [\frac{a(a+1)}{2}]^x$$

The number of possible genotypes with e.g. 100 variable loci, each with two alleles is 5 x 10^{47}! Crucial in the discussion of genetic variation is, however, how many loci are variable.

For a long time it was not possible to make quantitative statements on this matter and two contrasting views existed. The classical hypothesis assumed that most populations possessed very little variation due to the action of purifying selection. In this view adaptation to the prevailing environmental conditions would lead to genetic uniformity, because only the "best" genotype for a particular set of environmental variables would be maintained. Individuals would be homozygous at nearly all loci. Genetic variation in a species would only be found among populations adapted to different environments.

The opposite view, the balance hypothesis, supposed a lot of genetic variation, maintained by some kind of balancing selection as e.g. heterozygote superiority. Under this hypothesis individuals would be heterozygous at many loci.

Up to two decades ago it was virtually impossible to decide about the reality of the two models, as the techniques for random sampling of genetic variation were not available. Recent techniques, like DNA sequencing and RFLP assays, provide the means for quantitative screening of genetic variation (Hillis and Moritz, 1990; Hewitt *et al.* 1991).

Table 1. Mean fraction of polymorphic loci (P) and mean heterozygosity (H) based on surveys of allozyme variation in three groups of organisms. n is the number of species surveyed, standard errors in brackets (from [*] Hamrick and Godt 1990; and [+] Nevo *et al.* 1984).

	P	n	H	n
Plants [*]	0.342 (0.012)	468	0.113 (0.005)	468
Invertebrates [+]	0.375 (0.011)	371	0.100 (0.005)	361
Vertebrates [+]	0.226 (0.006)	596	0.054 (0.003)	551

Their large scale use, involving assays of ample numbers of individuals, has only just begun. Most of our insight with respect to the levels and distribution of genetic variation up to now comes from an older technique, protein electrophoresis. This method allows detection of genetic variation at the molecular level. As proteins have an electrostatic charge, due to some of the component amino acids, which are negatively or positively charged, they will migrate in an electrical field. After electrophoresis and staining, specific proteins can be identified. Genetic variation at the DNA-level for a particular protein can be detected by differences in electrophoretic migration distance, at least when the changes involve amino acid substitutions which cause differences in charge. Genetic differences at the DNA-level, generally caused by single base substitutions, can then be detected by electrophoresis. As only those base substitutions leading to differently charged protein molecules can be detected, the method provides an underestimation of the genetic variation present at the DNA-level. Because individual proteins can be identified by using specific staining techniques, genetic variation can be detected at separate loci. By screening large numbers of individuals for several loci, estimates of genetic variation can be obtained. Measures generally used are fractions of polymorphic loci (P) and heterozygosity (H). This method of screening a random sample of structural loci has been extremely useful for quantifying genetic variation. Many species have now been surveyed for this kind of genetic variation, called allozyme variation. Consequently our knowledge about the amount of variation and its distribution has been greatly enhanced.

        The general information obtained from allozyme studies is that populations of most animal and plant species are genetically highly variable. Table 1 gives a survey of allozyme variation in many species, derived from recent extensive studies of Nevo *et al.* (1984) and Hamrick and Godt (1990). Though nearly all species have considerable amounts of allozyme variation, differences among systematic groups occur. For example, insects generally have very high levels of allozyme variation and among plants,

Table 2. Examples of species in which no allozyme variation could be detected

Species	Number of loci	Reference
Mirounga angustirostris (northern elephant seal)	24	Bonnell & Selander (1974)
Acinonyx jubatus (cheetah)	52	O'Brien *et al.* (1985)
Phoca vitulina (harbour seal)	30	Swart & Van Delden (1994)

gymnosperms have high levels of allozyme variation. Birds and large carnivores on the other hand have low levels of allozyme variation. Table 2 lists a few species for which, despite a considerable number of loci screened, no allozyme variation could be detected. For the cheetah genetic monomorphism for the major histocompatibility complex was also found (O'Brien *et al.* 1985). The reasons for this lack of genetic variation are not clear, some possibilities are discussed below. Caution should be exercised in equating allozyme variation with other types of genetic variation, as e.g. variation for fitness characters. Comparison of levels of genetic variation for various characters are scarce and at present no general conclusions can be reached (Wolff, 1991). It is, however, generally agreed that for comparisons of genetic variation at a random sample of neutral genes, the use of allozyme variation is a reliable tool.

Forces acting on genetic variation

        Genetic variation present in populations is preserved when a number of conditions are fulfilled. The Hardy-Weinberg rule is the formal representation of this phenomenon. In addition it indicates that fixed genotypic proportions, based on allele frequencies, are to be expected. Most often, however, one or more conditions will not hold and changes in genetic composition will occur. Table 3 lists the effects on the level of genetic variation of the various forces acting in populations. Introduction of new genetic variants will occur through mutation and gene flow from genetically different populations. Assortative mating will mainly effect genotypic proportions.

Selection against (partly) recessives will eliminate alternative alleles and leave the population monomorphic at a particular locus. This is the selection model behind the classical model discussed above. In the case, however, that the heterozygote is the most fit genotype, a stable allele frequency equilibrium will be reached. Under this overdominance model, like in other forms of balancing selection such as in some cases of frequency dependent selection, genetic variation is protected. Loss of genetic variation will occur when genetic drift and inbreeding are predominant.

Table 3. Effects of various evolutionary forces on allele frequencies

	Variation within populations	Variation between populations
Mutation	+	−
Gene flow	+	−
Genetic drift/inbreeding	−	+
Directional selection	−	+/−
Balancing selection	+	−
Assortative mating	+/−	+/−

Genetic drift

Population genetic theory has provided models to analyze the above mentioned processes in a quantitative way. Here we will limit ourselves for the most part to genetic drift and inbreeding. The effects of these processes are most pronounced when the numbers of individuals in populations are low. Decline in population size is a situation that is being encountered progressively more often by many species. Hunting, toxification and habitat destruction are among the agents that limit population size. Populations of species, experiencing such detrimental conditions, often become small, fragmented and isolated. It is therefore relevant to consider the expected effects on genetic variation. Genetic drift will occur when population size is small. In a statistical sense it can be considered as taking a small sample from a large population. In terms of a genetically variable diploid population it means that out of a very large number of gametes produced, only a limited number will unite to form zygotes in the next generation. Because of the limited sample size, the allele frequencies of the individuals which form the new generation will often differ from those in the previous generation. This phenomenon is illustrated in Figure 1, for a number of subpopulations, consisting each generation of two individuals, and consequently having a total of four alleles at the A locus. Each subpopulation had initially started with 2A and 2a alleles, thus the initial allele frequency being 0.5. Due to genetic drift allele frequencies fluctuate in the course of generations and finally each subpopulation is fixed by chance for a particular allele and total homozygosity is reached. Genetic drift thus leads to loss of genetic variation in an isolated population. Considering the metapopulation consisting of all the test populations, it appears that the variance in allele frequencies among subpopulations will increase until fixation is complete. Then the expectation is that half of the subpopulations will be fixed for allele a and half for A, as the initial frequency was 0.5. An important observation at the genotypic level is that the proportion of homozygotes in the metapopulation will exceed the expectations based on the Hardy-Weinberg proportions associated with particular allele frequencies. In our example with a population size (N) of two, the effects of drift were extreme. In larger populations they will be more moderate in terms of the number of generations needed for complete fixation of subpopulations and the rate of increase in homozygosity of the metapopulation. The relation with population size is indicated in Figure 2. The chances of loss of an allele through drift are related with its frequency. Rare alleles will be readily lost. Figure 3 demonstrates the rapid loss of alleles

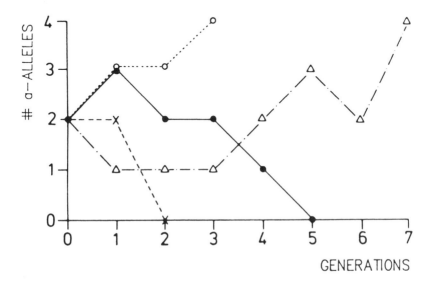

Figure 1. Genetic drift of allele frequency in four populations of a diploid species. Population size N = 2, initial allele frequency, 0.5.

in a 4 and a 12 allele case.

The impact of drift is strongly dependent on population size. Unfortunately, the actual population size, as determined in a natural population, is generally not indicative of the level of drift to be expected. Theoretical considerations show that for example bottlenecks in population size in the past have profound effects on the level of drift to be expected in a present population of considerable size. In order to determine the level of drift in such a situation, knowledge of the effective population size, N_e, is needed. In cases of fluctuations in population size, the relation between N_e and the population sizes $N_1, N_2, ... N_t$ in generations 1, 2, ... t respectively is given by:

$$\frac{1}{N_e} = \frac{1}{t} [\frac{1}{N_1} + ... + \frac{1}{N_t}].$$

This implies that N_e can be much smaller than N measured in a particular generation. Figure 4 illustrates the discrepancies between N and N_e for a particular case. Another reason that causes N_e to be smaller than N, is an unequal ratio of males and

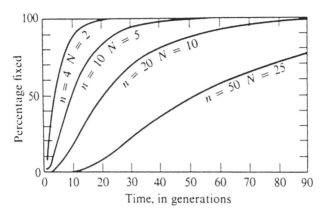

Figure 2 . The fraction of populations fixed by genetic drift in the course of generations at different values of N.

females contributing to the next generation. Such a situation will be encountered for example in animals with a harem system or in plants when only a limited fraction of individual plants produce pollen for the pollination of female plants. This will generally result in effective population sizes which are lower than the actual number of individuals, consequently the effects of genetic drift will be stronger than expected from the latter numbers.

Inbreeding

A process narrowly related to drift, and also predominantly associated with limited population sizes, is inbreeding. Inbreeding is defined as a higher incidence of consanguineous matings than expected in a model population. The level of inbreeding is given by the inbreeding coefficient, F, and is defined as the probability that two alleles at a locus in a diploid organism are copies of one and the same ancestral allele (Wright, 1969).

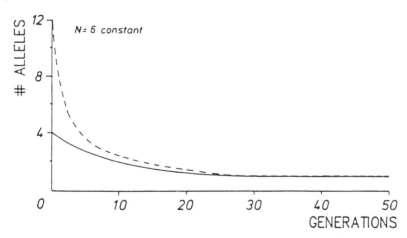

Figure 3. Loss of alleles in the course of generations by genetic drift in populations with constant size of six. The initial number of alleles were 4 and 12.

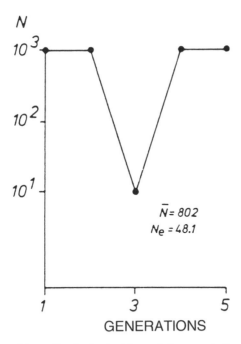

Figure 4. Example of the effect of a bottleneck in population size in one out of five generations. Mean population size (N) and effective population size (N_e) are indicated.

The relatedness of genetic drift and inbreeding is demonstrated by the use of F to quantify drift processes. The level of inbreeding in a population, indicated by F, is strongly determined by N. In a population of constant size, F will increase each generation. It can be shown that the increment of F, ΔF, is $1/(2N_e)$. The relation between F, population size and time (in generations) is shown in Figure 5. In inbreeding populations the fraction of heterozygotes is lower than expected under ideal Hardy Weinberg conditions. In a one-locus situation with two alleles, A and a, with frequencies respectively p and q, the fraction of AA homozygotes with inbreeding is $p^2(1-F) + pF$ and the fraction of Aa heterozygotes is $2pq(1-F)$, to be compared with respectively p^2 and $2pq$, under Hardy-Weinberg expectations.

Inbreeding depression

The increase in homozygosity with inbreeding is often accompanied by a phenomenon known as inbreeding depression. Inbred individuals have lower fitness than outbreds. Impressive data on the severity of inbreeding depression come from laboratory experiments (see Figure 6 as an example of inbreeding depression in mice) and from animal husbandry and plant breeding. Complementary to inbreeding is heterosis: when different inbred strains are crossed the F_1 is generally superior in fitness compared to the parental strains (Frankel 1983). Two theories are proposed for the explanation of inbreeding depression and heterosis. The partial dominance theory assumes that in populations of sexual outcrossing organisms, recessive deleterious alleles occur at many loci though in low frequency. Due to the increased homozygosity accompanying inbreed-

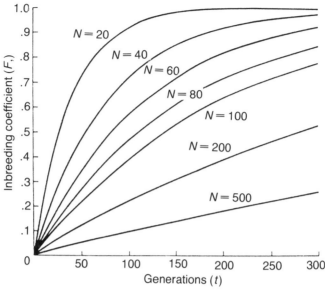

Figure 5. The increase in inbreeding coefficient F, in the course of generations at various values of N.

ing such alleles will become homozygous. Consequently the individuals involved will have reduced fitness. The presence of such deleterious alleles has been extensively shown in *Drosophila* species (review in Crow and Simmons, 1983) and humans (Cavalli Sforza and Bodmer, 1971). In this view heterosis arises from the combination in heterozygotes of different deleterious alleles which provide high fitness. In the overdominance theory it is assumed that heterozygotes are superior in fitness compared to homozygotes. Following inbreeding, heterozygosity will decline accompanied by fitness reduction. Restoration of heterozygosity as in hybrids between inbred lines, then explains the occurrence of heterosis.

In context with the previous section it should be mentioned that a growing set of data indicates the existence of a positive correlation between heterozygosity and fitness. Many studies on allozyme variation, both in animals and plants, prove that an individual that is heterozygous at many allozyme loci is often superior for particular fitness components like growth rate, production of offspring, survival and developmental homeostasis compared to less heterozygous individuals (Mitton and Grant 1984). This relationship, which is by no means universal, is hitherto unexplained. Though increased versatility of enzymes in allozyme heterozygotes may be an explanation, the level of allozyme heterozygosity may represent the general level of heterozygosity at all loci of the genome, including those with large fitness effects, undetected in electrophoretic studies. In view of the heterozygosity-fitness relation it has been proposed that a high level of heterozygosity would enable a population or species to exploit a wider range of environments. At the species level this could mean that among related species, those with higher levels of heterozygosity would occur in larger, more variable, habitats than species with low levels of heterozygosity. Van Valen (1965) has advocated this view in his niche width hypothesis. Evidence for this hypothesis comes from comparisons of narrowly related plant species, differing in heterozygosity (Babbel and Selander, 1974; Karson, 1987).

Breeding system

The breeding system of a species has a profound effect on the degree of inbreeding and consequently on the level and organization of genetic variation. Figure 7 shows the increase of F in the course of generations with various breeding systems. Especially in self-fertilizing organisms inbreeding levels will increase very rapidly. Still many plant species exhibit high levels of selfing and it may be asked which consequences with respect to inbreeding depression are to be expected in such cases. As predominantly selfing species seem to flourish equally well as obligate outbreeders, apparently no or only moderate deleterious effects occur. The probable reason for this is that detrimental

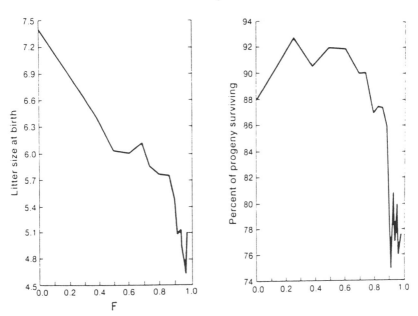

Figure 6 . Inbreeding depression for two fitness components in the house mouse at various values of F.

alleles in selfing species have been eliminated in the past (Lande and Schemske 1985; Charlesworth *et al.* 1990). A recent study on the effects of inbreeding in two populations of *Eichhornia paniculata*, one predominantly outcrossing, the other predominantly selfing, is in conformity with this view (Barrett and Charlesworth, 1991). On the other hand, many hermaphrodite plant species possess self-incompatibility systems, which prevent selfing. This is commonly seen as a mechanism to prevent the deleterious effects of selfing. In studies on the organization of genetic variation and on inbreeding it is therefore of great importance to know the breeding system of the species.

Extinction and genetic variation

Species, which are exposed to habitat destruction or habitat fragmentation, will often consist of isolated populations of limited population size. Such small populations have a relatively high chance of becoming extinct due to purely demographic reasons. These populations generally carry a higher risk of attaining a zero population size due to fluctuations in size than large populations. The genetic processes in small populations

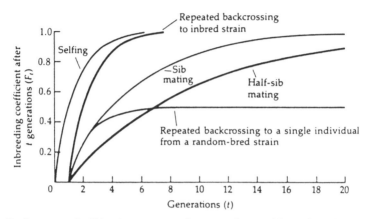

Figure 7. Increase in F in the course of generations with various breeding systems.

previously described may also contribute to an increased risk of extinction. The lower individual fitness to be expected in such populations may augment the vulnerability to current and future stresses. In this respect two aspects of genetic drift and inbreeding are highly relevant. The first is concerned with the loss of specific alleles at one or at a restricted number of loci. When such variants provide individuals with tolerance to specific environmental stresses, their loss will lead to deterioration of individual adaptedness. A population consisting of individuals which have lost the potency for adaptation will be more vulnerable to extinction when exposed to a specific stress. The second aspect implies the general loss of fitness associated with inbreeding depression. In this case the probability of extinction is not limited to one particular stress factor but will occur more generally, though the effects of inbreeding depression will generally be more severe under harsh conditions.

With respect to the first mentioned case, the relation of particular genetic variants to specific stresses, numerous examples are known where adaptation has been strongly increased in natural populations. These examples include classical cases like industrial melanism in moths (Kettlewell 1973; Brakefield 1987), sickle cell polymorphism in humans (Allison 1964), warfarin resistance in rats (Greaves et al. 1977), insecticide resistance in insects (Roush and McKenzie 1987) and heavy metal tolerance in plants (Baker 1987). It may be inferred from the gain in individual fitness, due to the presence of particular genetic variants, that the populations involved have acquired also increased population fitness. The latter concept is much less well defined than individual fitness and is often disputed (Lewontin 1970). Short-term population fitness may be defined as the existence of a population over a limited number of generations. Fitness differences among populations are then reflected in differential survival of populations or in differences in population size. The concept of long-term fitness is applied for much longer periods of time, during which considerable changes in environmental conditions may occur. Thoday (1953) has defined long-term population fitness as the "probability that a contemporary group of individuals will survive for a given long period of time, such as 10^8 years, that is to say will leave descendants after the lapse of that time". In this context Lewontin (1957, 1961) has proposed an all-or-none fitness concept: if a population can survive and reproduce in more environments than another population, the former possesses a greater fitness. It can then be argued that a population which survives in a greater number of environments than others also has a greater chance to survive future environments. This enables an experimental approach of long-term fitness by testing populations in a great number of different environments.

The importance of the presence or absence of particular genetic variants is demonstrated by an experiment in which populations of *Drosophila melanogaster* differing in genetic composition for the alcohol dehydrogenase (*Adh*) locus, were submitted to various kinds of stress (Bijlsma-Meeles and Van Delden 1974). The *Adh* locus is polymorphic for two alleles *Fast* (*F*) and *Slow* (*S*). The *Adh* genotypes differ considerably in survival on medium containing toxic concentrations of alcohols. The *F*-homozygote survives more often than the *S*-homozygote, while the heterozygote is intermediate in survival. Survival is positively correlated with ADH-activities of the genotypes (Van Delden, 1982). When large numbers of vial populations either monomorphic for the *S*-allele, monomorphic for the *F*-allele or polymorphic were continuously exposed to toxic concentrations of ethanol, some populations did not survive in the course of several generations. However, extinction fractions differed considerably among population types (Figure 8). Monomorphic *S*-populations had higher extinction rates than the other population types, apparently because they missed the *F*-allele, which provides higher individual fitness to its carriers when exposed to ethanol. Individual fitness differences are thus transferred to the population level and result in differential extinction probabilities. In a set of additional experiments populations were exposed to varying stress factors in successive generations (Van Delden, unpublished). The populations were exposed during one generation to one particular stress condition (like high temperature, low humidity, ethanol addition to the food, etc.) and in the next generation to another factor (randomly chosen from a set of conditions) and so on. Again differential extinction was observed. In this case, however, the polymorphic populations, has the lowest extinction rates. This experiment points to the importance of polymorphism in varying environments.

The other potentially important cause for increased rates of extinction is the general reduction in genetic variation following inbreeding, associated with inbreeding depression. Examples of the deleterious effects of inbreeding depression are for example the data from O'Brien *et al.* (1985) for several species of mammals kept in captivity. Juvenile mortality is one of the most prominent traits effected by inbreeding depression in these species. It has been shown for *Drosophila melanogaster* that population size and productivity are severely reduced following inbreeding (Van Delden and Beardmore 1968; Van Delden unpublished). When genetic variation in such populations was artificially increased by X-irradiation considerable improvement was obtained. Merely the reduction in population size will increase the probability of extinction of local, isolated populations. It is a well known fact that when *Drosophila* lines are inbred, a considerable number of lines are lost in the course of generations. Some of these extinctions will occur because of homozygosity for recessive lethals, others will be due to the increased chance of extinction associated with reduced population size. Falconer (1960) has made the same observation in inbred strains of mice. Litter size decreased with about 0.53 young per 10% increase in F. By the time F had reached 0.76 only 3 out of 20 lines survived.

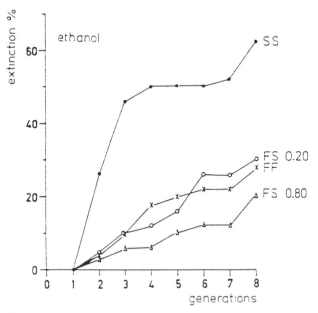

Figure 8. Extinction percentages of *Drosophila melanogaster* populations, either monomorphic or polymorphic (either started with an initial *Adh* frequency of 0.20 or with a frequency of 0.80).

Prospects with respect to natural populations

Though we have well established expectations of the relations between population size, levels of genetic variation and fitness based on theoretical considerations, laboratory experiments and data from animal and plant breeding, the significance for natural populations is still largely unclear. In a recent review on inbreeding depression, however, Charlesworth and Charlesworth (1987) in considering the available knowledge, hold the view that also in nature, prevailing conditions will often lead to inbreeding depression. Among studies in this field we may cite the investigation of Quattro and Vrijenhoek (1989) on three populations of the endangered fish species *Poecillopsis occidentalis* from Arizona, USA, differing in levels of allozyme variation. A positive correlation was found between heterozygosity and four fitness components.

Studies in this field are comprehensive and time-consuming because information is needed on the demography of populations (both in the present and the past), as well as on levels of genetic variation and fitness. Such a study has been undertaken for the plant species *Salvia pratensis* and *Scabiosa columbaria* (Van Treuren *et al.* 1991; Ouborg *et al.* 1991; Bijlsma *et al.* 1991). Both species experienced during the last decades a sharp decline in the number of populations occurring in the Netherlands, while part of the remaining populations are reduced to low numbers of individuals. Plants were sampled from populations with varying population sizes and screened for allozyme variation. Significant positive correlations were found between population size and both proportion of polymorphic loci and observed number of alleles. A positive correlation was also found between population size and amount of phenotypic variation for a number of morphological characters. Part of this variation had a genetic origin. These findings thus confirm the expectation with respect to the relation between amounts of genetic variation and population size. In an experiment with six populations of *Scabiosa colum-*

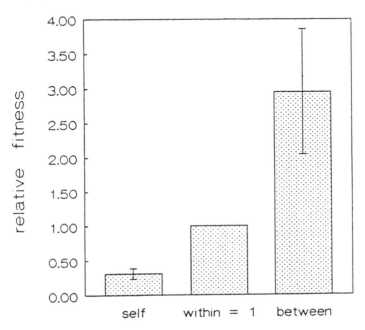

Figure 9. Fitness differences in *Scabiosa columbaria* in F_1's from selfing, crosses within and between populations (explanation in text).

baria, varying in population size from 35 - 100,000 plants, three kinds of crosses were made: plants were selfed, crosses were made between individuals of the same population and crosses were made between individuals of different populations. The offspring of the three kinds of crosses were tested in several phases of the life cycle for a number of fitness components like survival, biomass production and seed production. The experiment included stress conditions like competition and cutting. Invariably the progeny of selfed plants showed dramatically lower fitness than the progeny of the outcrossed groups, while the progeny of the between-population crosses generally had higher fitness than the progeny of the intra-population crosses (Figure 9). The experiment thus demonstrates the detrimental effects of inbreeding, which were most distinct under stress conditions. The effects of population size were ambiguous, however. This may result from the way population sizes were estimated, which occurred by simply counting the number of adult plants present at the time of the experiment. Such a survey does not take into account factors like past population sizes, size of seedbanks, etc. Plant density, more than population size, seems to effect inbreeding depression in this insect-pollinated species (Van Treuren unpublished).

Studies on the relation between population demography, genetic variation and fitness are of great interest from an evolutionary viewpoint. Especially the role of the breeding system in determining the genetic structure of populations is a fascinating object of study. In an era where many species experience progressive habitat destruction and consequent reduction in numbers of individuals together with fragmentation and isolation of populations the predicted loss of genetic variation may have profound effects. In the last decade therefore considerable interest has been shown for the relation between population size, genetic variation and extinction (Frankel and Soulé 1981; Schonewald-Cox *et al.* 1983; Soulé 1986; Soulé 1987; Seitz and Loeschcke 1991). Our knowledge is, however, still limited and further theoretical and experimental work is needed. More information is needed on the effects of bottlenecks in population size on genetic variance for quantitative characters. It has been shown that in some cases additive genetic variances increased as a result of a bottleneck (Bryant *et al.* 1986), maybe due to disrupted covariance matrices (Carson 1990). Also the relation between population extinction, genetic variation and demographic stochasticity has hardly been explored (Gabriel *et al.* 1991). Further important items are the effects of gene flow and the concept of the metapopulation level in relation to demographic and genetic equilibrium (Olivieri *et al.* 1990). Therefore, in view of its potential relevance for conservation studies and the lack of data from natural populations, the study of the relation between genetic pauperization and population- or species-extinction should be intensified.

Acknowledgements

I thank R. van Treuren for permission to use some of his unpublished results and R. Bijlsma and R. van Treuren for stimulating discussions.

References

Allison, A.C., 1964. Polymorphism and natural selection in human populations. *Cold Spring Harbor Symp. Quant. Biol.* 29: 137-149.

Babbel, G.R. and R.K. Selander, 1974. Genetic variability in edaphically restricted and widespread plant species. *Evolution* 28: 619-630.

Baker, A.J.M., 1987. Metal tolerance. *New Phytol. (suppl.)* 106: 93-111.

Barrett, S.C.H. and D. Charlesworth 1991. Effects of a change in the level of inbreeding on the genetic load. *Nature* 352: 522-524.

Bijlsma, R., N.J. Ouborg and R. van Treuren, 1991. Genetic and phenotypic variation in relation to population size in two plant species: *Salvia pratensis* and *Scabiosa columbaria*. In A. Seitz and V. Loeschcke (eds.), *Species conservation: a population-biological approach.* pp. 89-101. Basel: Birkhauser Verlag.

Bijlsma-Meeles, E. and W. van Delden, 1974. Intra- and inter-population selection concerning the Alcohol Dehydrogenase locus in *Drosophila melanogaster. Nature* 247: 369-371.

Bonnell, M.L. and R.K. Selander, 1974. Elephant seals: genetic variation and near extinction. *Science* 184: 908-909.

Brakefield, P.M., 1987. Industrial melanism: do we have the answers? *Trends Ecol. Evol.* 2: 117-122.

Bryant, E.H., S.A. McCommas and L.M. Combs, 1986. The effect of an experimental bottleneck upon quantitative genetic variation in the housefly. *Genetics* 114: 1191-1211.

Carson, H.L., 1990. Increased genetic variance after a population bottleneck. *Trends Ecol. Evol.* 5: 228-230.

Cavalli-Sforza, L.L. and W.F. Bodmer, 1971. *The genetics of human populations*. San Francisco: Freeman.

Charlesworth, D. and B. Charlesworth, 1987. Inbreeding depression and its evolutionary consequences. *Ann. Rev. Ecol. Syst.* 18: 237-268.

Charlesworth, D., M. T. Morgan and B. Charlesworth, 1990. Inbreeding depression, genetic load, and the evolution of outcrossing rates in a multilocus system with no linkage. *Evolution* 44: 1469-1489.

Crow, J.F. and M.J. Simons, 1983. The mutation load in *Drosophila*. In M. Ashburner, H.L. Carson and J.N. Thompson (eds.), *The genetics and biology of Drosophila*, vol. 3c, pp. 1-35. London: Academic Press.

Falconer, D.S., 1960. The genetics of litter size in mice. *J. Cell. Comp. Physiol.* 56: 153-167.

Frankel, O.H. and M.E. Soulé, 1981. *Conservation and evolution*. Cambridge: Cambridge University Press.

Frankel, R. (ed.), 1983. *Heterosis: reappraisal of theory and praxis*. Berlin: Springer Verlag.

Gabriel, W., R. Bürger and M. Lynch, 1991. Population extinction by mutational load and demographic stochasticity. In A. Seitz and V. Loeschcke (eds.), *Species conservation: a population biological approach*, pp. 49-59. Basel: Birkhaüser Verlag

Greaves, J.H., R. Redfern, P.B. Ayres and J.E. Gill, 1977. Warfarin resistance: a balanced polymosphism in the Norway rat. *Genet. Res.* 30: 257-263.

Hamrick, J.L. and M.J.W. Godt, 1990. Allozyme diversity in plant species. In A. H. D. Brown, M. T. Clegg, A. L. Kahler and B. S. Weir (eds.), *Plant population genetics, breeding and genetic resources*, pp. 43-63. Sunderland, Mass.: Sinauer.

Hewitt, G. M., A. W. B. Johnston and J. P. W. Young (eds.), 1991. *Molecular techniques in taxonomy*. NATO ASI Series H, Vol. 57. Berlin: Springer Verlag.

Hillis, D. M. and C. Moritz (eds.), 1990. *Molecular systematics*. Sunderland, Mass.: Sinauer.

Karron, J. D. 1987. A comparison of levels of genetic polymorphism and self-compatibility in geographically restricted and widespread plant congeners. *Evolutionary Ecology* 1: 47-58.

Kettlewell, H. B. D. 1973. *The evolution of melanism*. Oxford: Clarendon Press.

Lewontin, R. C. 1957. The adaptation of populations to varying environments. *Cold Spring Harbor Symp. Quant. Biol.* 22: 395-408.

Lewontin, R. C. 1961. Evolution and the theory of games. *J. Theor. Biol.* 1: 382-403.

Lewontin, R.C. 1970. The units of selection. *Ann. Rev. Ecol. Syst.* 1: 1-18.

Nevo, E., A. Beiles and R. Ben-Slomo, 1984. The evolutionary significance of genetic diversity: ecological, demographic, and life history correlates. In G. S. Mani (ed.), *Evolutionary dynamics of genetic diversity*, pp. 13-213. Berlin: Springer Verlag.

O'Brien, S.J., M.E. Roelke, L. Marker, A. Newman, C.A. Winkler, D. Meltzer, L. Colly, J.F. Evermann, M. Bush & D.E. Wildt, 1985. Genetic basis for species vulnerability in the cheetah. *Science* 227: 1428-1434.

Olivieri, I., D. Couvet and P.H. Gouyon, 1990. The genetics of transient populations: research at the metapopulation level. *Trends Ecol. Evol.* 5: 207-210.

Ouborg, N. J., R. van Treuren and J. M. M. van Damme, 1991. The significance of genetic erosion in the process of extinction. II. Morphological variation and fitness components in populations of varying size of *Salvia pratensis* L. and *Scabiosa columbaria* L. *Oecologia* 86: 359-367.

Quattro, J.M. and R.C. Vrijenhoek, 1989. Fitness differences among remnant populations of the endangered Sonoran Topminow. *Science* 245: 976-978.

Roush, R. T. and J. A. McKenzie, 1987. Ecological genetics of insecticide and acaricide resistance. *Ann. Rev. Entomol.* 32: 163-179.

Schonewald-Cox, C.M., S.M. Chambers, B. MacBride and W.L. Thomas (eds.), 1983. *Genetics and conservation*. Menlo Park: Benjamin/Cummings.

Seitz, A. and V. Loeschcke (eds.), 1991. *Species conservation: a population-biological approach*. Basel: Birkhäuser Verlag.

Soulé, M.E. (ed.), 1986. *Conservation Biology. The science of scarcity and diversity*. Sunderland, Mass.: Sinauer.

Soulé, M.E. (ed.), 1987. *Viable populations for conservation*. Sunderland, Mass.: Sinauer.

Swart, S. and W. van Delden, 1994. Lack of genetic variation in common seals. *Heredity* (in press).

Thoday, J.M., 1953. Components of fitness. *Symp. Soc. Exp. Biol.* 7: 96-113.

Van Delden, W., 1982. The alcohol dehydrogenase polymorphism in *Drosophila melanogaster*. *Evol. Biol.* 15: 187-222.

Van Delden, W. and J. A. Beardmore, 1968. Effects of small increments of genetic variability in inbred populations of *Drosophila melanogaster*. *Mutation Res.* 6: 117-127.

Van Treuren, R., R. Bijlsma, W. van Delden and N.J. Ouborg, 1991. The significance of genetic erosion in the process of extinction. I. Genetic differentiation in *Salvia pratensis* and *Scabiosa columbaria* in relation to population size. *Heredity* 66: 181-189.

Van Valen, L. 1965. Morphological variation and width of ecological niche. *Amer. Nat.* 99: 377-390.

Wolff, K. 1991. Analysis of allozyme variability in three *Plantago* species and a comparison to morphological variability. *Theor. Appl. Genet.* 81: 119-126.

Wright, S. 1969. *Evolution and the genetics of populations*, volume 2. Chicago: University of Chicago Press.

7. Geophysiological aspects of biodiversity

James E. Lovelock

Alfred Lotka (1925), in his book 'The Elements of Physical Biology', warned that the mathematical understanding of the evolution of organisms would be difficult if the evolution of their physical environment was not considered simultaneously. This warning has been prophetic and Robert May (1981) has shown that attempts to model the evolution of natural ecosystems from the bottom up through population biology alone, are doomed to fail as a result of inherent mathematical instability. Population biology seems limited to simple models containing one or two species only, or to highly damped density dependent systems.

The object of this paper is to present a different kind of population biology model, one that appears to be almost entirely free of the limitations just mentioned. The new model will accommodate at least hundreds of species and uses natural non-linear growth equations. Its inherent stability permits model experiments with multispecies ecosystems including several trophic levels. It is stable against perturbations and almost entirely insensitive to the initial conditions. If this new model can be taken as a fair caricature of the natural world then it seems to provide new insights into biodiversity.

Geophysiology is the basis of this model. Geophysiology is a reanimation of the father of geology, James Hutton's view of the Earth as something behaving like a physiological system. It is a top down view of planetary processes that sees the evolution of organisms and the evolution of their chemical and physical environment to be so tightly coupled as to behave as a single entity. This way of looking at planetary life is to be contrasted with the usual, bottom up, view that sees the evolution of organisms as separate or only loosely coupled with the evolution of their material environment. The main difference between these two theoretical views of the Earth is that geophysiology includes in its models the natural physical and chemical constraints to growth that automatically prevent unrestrained instability.

The geophysiological model presented in this paper was made in an attempt to answer the question: is the Earth to any extent self regulating with respect to climate or chemical composition?

I wondered if a component of the biosphere, through its growth and competition for space, or resources, could interact with the environment in a way that made the entire system self regulating.

I chose as a model a very simple system, one where the organisms were represented by a single species of plant, daisies, and the environment by a single variable, temperature. A reductionist approach maybe, but biologists, recognizing the complexity of their subject, do not usually object to reduction. The model which I call, Daisyworld, was stable in operation and resistant to perturbations. As a population biology model, it was unusual in its ability to accept almost any initial conditions, and, like a stable engineering control system, worked immediately it was switched on.

The model itself and its use in experiments on biodiversity is described below. The mathematical basis, the equations used are described in Lovelock (1981), Lovelock and Watson (1983), and Maddock (1991). Before discussing the model, and the biodiversity experiments made with it, I need to comment briefly on its scientific validity and show that Daisyworld is more than just a flight of fancy, or a hollow Trojan horse bearing a new and dangerous form of animism. The model is broadly accepted by climatologists and its combination of simplicity and stability has led to its use by Henderson-Sellers and McGuffie (1987), to demonstrate for teaching purposes, the regulation of climate through changes in the planetary albedo. Zeng et al (1990) tried to demolish it by inserting a time lag between the sensing of heat input and the response of the system. Any engineer or physiologist would have told them that such an act is a recipe for instability and chaos. Jascourt and Raymond (1991) subjected the Daisyworld model to an extensive mathematical analysis and reported it to be soundly based. Perhaps the most intriguing adaptation of the model is as part of a computer game called SimEarth. As in climatology teaching, Daisyworld provided a stable and well behaved platform on which to build more intricate model scenarios. I hope that Daisyworld will be judged by its likeness as a caricature of the real world.

Description of Daisyworld

To understand Daisyworld, imagine a planet, like the Earth but with less ocean, orbiting a star like the Sun. The star is assumed to increase its output of heat and light as it ages, just as our own Sun is thought to have done. The planetary surface is everywhere well watered and fertile and is sown with the seeds of two daisy species, one colored dark and the other light. The object of the model is to demonstrate that, as the star heats up, the growth and competition for space between species of daisies can keep the planetary temperature always comfortable for plants. This model was first described by Lovelock (1983) and subsequently in more detail by Watson and Lovelock (1984). These papers and that of Maddock (1991) list the equations used and the methods for their integration.

Figure 1 shows the conventional wisdom about the rise of temperature and the growth of organisms on Daisyworld as the star evolves and grows hotter. The upper panel shows the smooth monotonic rise of temperature of the planet according to the Stefan-Boltzmann relationship and the lower panel the parabolic rise and fall of daisy growth as the planet warms from 5 degrees Celsius, the lower limit of temperature for growth, through comfortable temperatures to the upper limit of temperature for growth, 40 degrees Celsius. The two panels of this diagram represent the geophysical and biological views of the evolution of climate, and of the growth of organisms on the

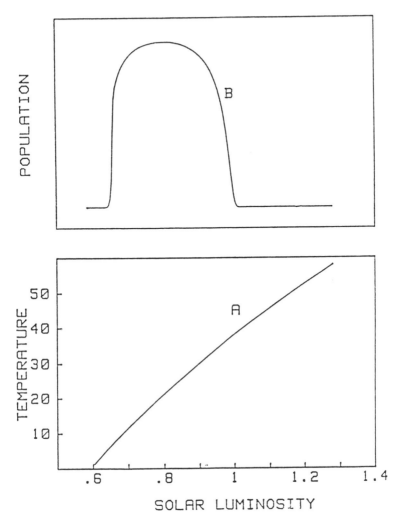

Figure 1. Models of the evolution of Daisyworld according to conventional wisdom. The top panel (B) illustrates daisy population in arbitrary units. The bottom panel (A) shows temperature in degrees Celsius. Going from left to right along the horizontal axis the star's luminosity increases from 60 to 140% that of our Sun. The two evolutionary processes are treated quite separatedly by biologists and physicists.

imaginary planet, and are what would happen if the growth of organisms was not tightly coupled to the physical evolution of the environment.

Figure 2 shows how in fact Daisyworld does evolve. At the start of the first season after the planetary temperature reached 5 degrees Celsius, daisy seeds would begin to germinate. After their emergence, dark colored daisies would be at an advantage since in the feeble sunlight they alone would be warm enough to grow. The few seeds left at the end of season would nearly all be of dark daisies. At the start of the next season dark daisies would dominate and soon begin to spread, warming themselves and the area they occupied. Then, with explosive positive feedback, temperature and daisy growth

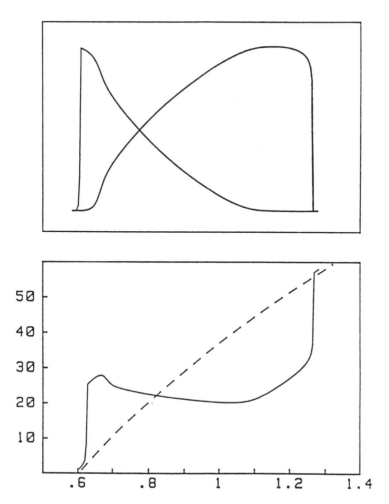

Figure 2. Models of the evolution of Diasyworld according to geophysiology. The two panels and the horizontal axis of the figures are the same as for Figure 1 but here the two evolutionary processes are tightly coupled.

would rise until a large proportion of the planetary surface was covered by dark daisies. Their growth would not continue indefinitely for two reasons: firstly, too high a temperature suppresses growth, and secondly, on a warm planet there would be competition for space from light colored daisies. As the star warmed, the planetary ecosystem would change from one dominated by dark daisies to domination by light colored daisies. It is the nature of stars to grow hotter as they age and eventually the ecosystem of daisies would collapse when a total planetary surface cover of light daisies was insufficient to keep the planet cool.

Figure 3 illustrates how the system works. The inverted U shaped curve delineates the limits to the growth of daisies with temperature. The solid curved line illustrates the planetary mean temperature for different extents of cover with one light colored daisy species. These two curves intersect at the point where the system settles

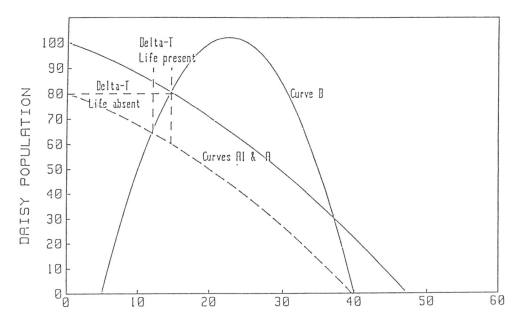

Figure 3. The mechanisms of Daisyworld. The inverted parabolic curve B shows the variation of daisy growth rate with temperature. The two convex curves A and A1 show the response of the planetary temperature to the area covered with light colored daisies. The intersection points of these curves on the left hand side of the diagram mark the dynamically stable states of population and temperature. Note the horizontal difference between the two convex curves. These differences represent the temperature changes for a decrease in solar output with and without regulation.

down to its dynamic equilibrium for temperature and daisy population. Imagine the star to suddenly decrease its output of heat. The resulting new equilibrium is indicated by the point of intersection of the dashed line with the growth curve. If there were no self regulation the temperature decrease would simply be the horizontal distance between the two dashed lines. With self regulation the temperature change is the horizontal distance between the two intersection points. The diagram illustrates how a system carrying but one species of daisy is able greatly to reduce the fall of temperature due to the decrease of solar output.

The model is quite general and works as well if the growth of the organisms alters the cloud cover, or the abundance of greenhouse gases. Regulation is not limited to temperature, and can be modelled for other environmental properties, like redox potential, or the abundance of atmospheric gases (Lovelock 1989). The model is robust in a mathematical sense and uses equations in their natural non-linear form and is almost entirely insensitive to the initial conditions.

At meetings where I have described Daisyworld a frequent criticism is that the model works only when the species are artificially chosen and that on a real planet there would always be cheats, organisms that took advantage of the small energy gain of not

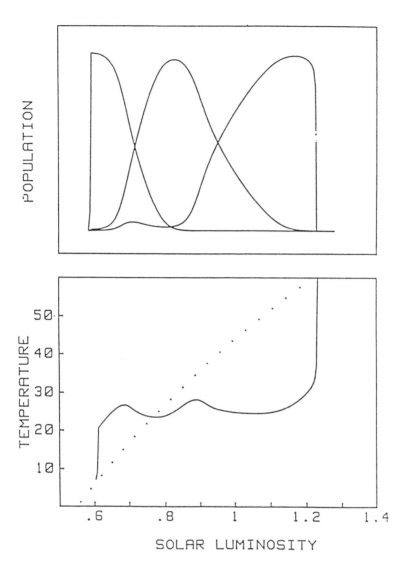

Figure 4. The evolution of climate on a three species Daisyworld with dark, neutral and light colored daisies present. The temperature evolution in the basence of life is shown for comparison as a gently curving convex line.

making pigment. Their growth they said would overwhelm the simple system of Daisyworld.

Figure 4 illustrates the effect of adding a third neutral colored specie to Daisyworld, one that occupies space without contributing to regulation, and which is given an artificial 5% bonus of growth rate for not making pigment. The three species Daisyworld is as stable and able to regulate as is the two species model. Cheating does not happen on Daisyworld because only dark daisies are fit to grow at low temperatures and only light daisies fit to grow at high temperatures. The growth of the neutral daisies or other organisms is restricted to that region where regulation is not needed.

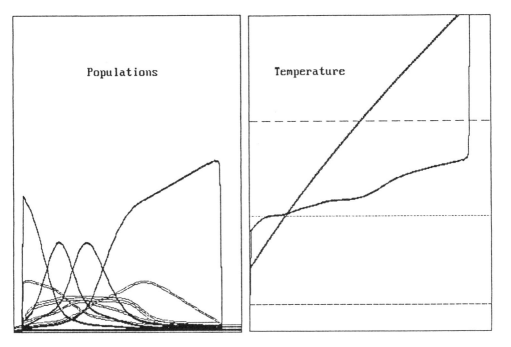

Figure 5. A Daisyworld populated by four different colored daisies, three different colored rabbits and a fox. The left hand panel shows the species populations and the right hand panel the evolution of the climate as the star warmed from 60 to 130% that of our Sun.

When I made this three species model I was only vaguely aware of the limitations of population biology models (Nicolis 1991; May 1976). I did not then know about the limitation to two species or about chaos. When I did read about the difficulties of modelling multispecies systems I felt for a while as if I were the scientific equivalent of Moliere's *Bourgeois Gentilhomme* who never knew that what he wrote was prose. It might seem that the great stability of Daisyworld is in contradiction to the inherent instability leading to deterministic chaos, described by Nicolis (1991) as a general property to evolving systems of non-linear differential equations. It would be wrong so to conclude, Daisyworld is more like an island of stability in a sea of chaos, the exception that proves the rule.

The damping effects of population density are well known to confer stability on both natural and model ecosystems (Hassel 1975; May 1981). What we have in Daisyworld is something quite different, a tightly coupled system made stable by environmental feedback. In Daisyworld unlimited growth does occur, but only when positive feedback is needed to bring the system rapidly to its stable state. This is illustrated in figure 3, by the rapid rise of temperature and daisy population at the origin of Daisyworld. It may be some time before evidence and observation confirms or denies the existence of systems like Daisyworld on the Earth, although it is already possible to argue, from the geochemistry of rock weathering, that this mechanism operates for the long term regulation of atmospheric carbon dioxide and climate (Lovelock and Watson 1983). For

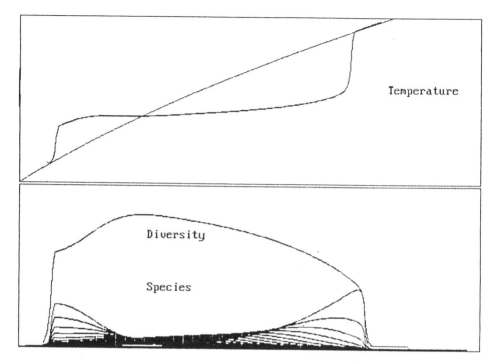

Figure 6. A Daisyworld with 20 different colored daisies. The upper panel shows the climate evolution as the star warmed from a luminosity of 60 to 130%. The lower panel shows the populations of the different colored daisies, diversity index and the total biomass. In this model only two integration cycles took place for each of the 200 temperature increments going from left to right along the horizontal axis.

the moment let us assume it to be a fair model and see how species richness and diversity can be examined on this imaginary planet.

Species richness and diversity in Daisyworld

First, there seems to be no limit to the number of species that can be accommodated in Daisyworld models. Figure 5 illustrates the evolution of climate and population on a world inhabited by four species of colored plants, three species of herbivores, and one carnivore. During its evolution this system is perturbed continuously by a progressive growth of solar luminosity and by a step increase in luminosity of 4 per cent.

It takes a robust system to continue to regulate and remain stable when so perturbed. The peaceful coexistence of daisies, rabbits and foxes is remarkable in itself, even without the perturbations.

The model I have most used to examine biodiversity is a multispecies Daisyworld. I used the Shannon index of biodiversity:

$$H = \sum (p_i) \ln (p_i)$$

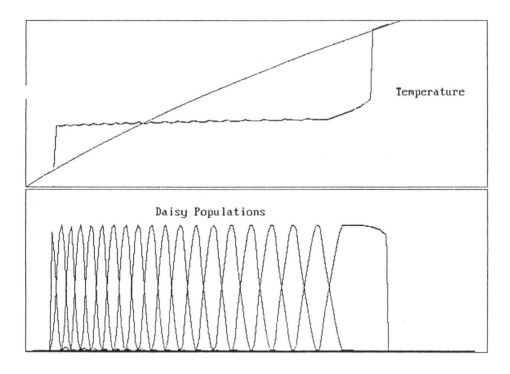

Figure 7. The model exactly as in Figure 6 but with 9 999 integration cycles for each temperature increment. Note the presence of only one or two daisy species at any one time.

I share Robert May's (1981) instinct to prefer some function of the variance as an index of biodiversity, but started my models with the Shannon index as the first I encountered. For this series of models the conclusions do not seem to depend significantly on which index is used.

Figure 6 shows the evolution of the climate, the populations of 20 species of daisies, differing only in the shade of their color, and the diversity index. Biodiversity appears to be greatest when the solar output is such that the temperature would be comfortable even with no regulation taking place. Diversity is least when the system is furthest from this comfortable zone, whether too hot or too cold. Not so different perhaps from the well known decrease of diversity with altitude or latitude.

My colleague Linda Maddock (1991) observed that most discussions of Daisyworld refer to conditions where the change of solar heat input was rapid, that is changing significantly during the generation time of the organisms. In other words, the system was always far from equilibrium. She observed that when the models were run for long periods at constant solar input, the number of species slowly declined until there were, at any time, never more than two species present. I failed to notice this in my earlier experiments with Daisyworld because my interest was in climate regulation, not biodiversity; moreover, I had noticed that changing the number of species initially available to

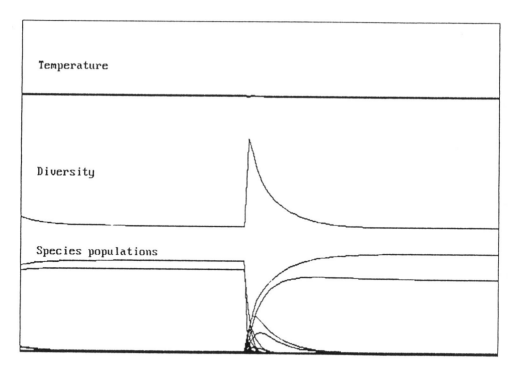

Figure 8. The perturbation of a 100 species Daisyworld carrying daisies that differed in color by evenly spaced intervals from an albedo of 0.25 to 0.65. The system was run through several million integration cycles until close to equilibrium and then perturbed by a 4% increase in solar luminosity. Note the large increase of the diversity index that accompanied a barely perceptible rise in temperature.

the model, had only a slight effect on the ability of the system to regulate. Looking more closely at both regulation and biodiversity I now see that although thermostasis is not determined by the number of species present at any given time, it is somewhat dependent upon the potential of the system to provide species when needed.

Figure 7 is of the same model as that illustrated in Figure 6, but here the rate of change of heat input from the star was so slow that the daisies were always in equilibrium with their environment. This required 10,000 integration steps for each of the 200 temperature increments along the time axis of the figure.

With these conditions the figure shows that there were never more than two species present at any time. Regulation of temperature was as effective as in figure 6 when the system evolved rapidly and many species were present simultaneously.

Figure 8 illustrates the effects on diversity in a 100 daisy species model of a step increase of luminosity of 4%, similar to the change in insolation experienced by the Earth as it moves from the glacial to the interglacial epochs. Before the change the system was at equilibrium and had settled down with only two species keeping a comfortable regime. The small sudden heat increase is seen to induce a burst of speciation followed by a slow return to equilibrium at the higher solar input.

The model illustrated in figure 5 has three trophic levels compared with the single level of Daisyworld. If this model is run to equilibrium at constant solar input the number of species present is between 3 and 5. It would seem that a minimum of one and a maximum of two is the equilibrium number of species for each trophic level of this type of model, and the total number for equilibrium of the system the simple addition. More complex models where several environmental variables are modelled simultaneously have been tried and found stable (Lovelock 1989). The next step is to see what effect increasing the complexity of the environment has on species richness.

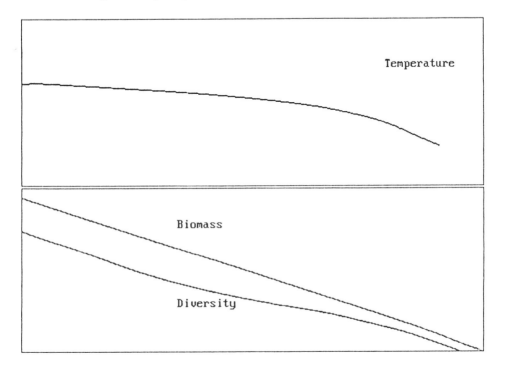

Increasing Death Rate -->

Figure 9. The effect of a progressively increasing death rate on a 20 species Daisyworld when the solar luminosity is constant at 0.7. The upper panel illustrates temperature regulation and the lower panel biomass and diversity index. The death rate was taken to increase from 0.05 to 1.0 along the horizontal axis.

Figure 9 illustrates the effect of an increase in the natural death rate on biodiversity. Here a system of ten daisies was run at constant solar input but with an increasing rate of natural death, such as by a progressive increase in abundance of some toxic material. The model shows the persistence of temperature regulation until almost all the daisies had died; the total population and the diversity index are also shown and both decline as the model develops. It seems that pathological stress, as by an increase of death rate, reduces biodiversity, whereas environmental stress, a change in temperature, increases biodiversity, if it is potentially available.

Figure 10 summarizes the results of this series of experiments with Daisyworld and shows how the diversity index varies with the logarithm of the rate of change of heat

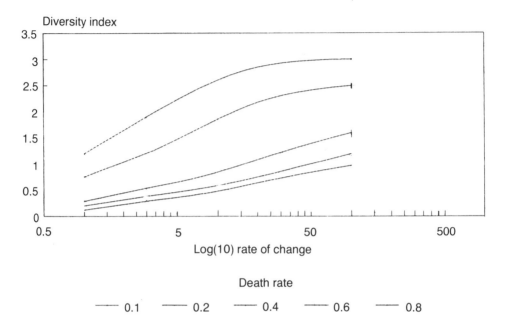

Figure 10. The relationship between the Shannon diversity index and the logarithm of the rate of change of solar luminosity going from 1.0 to 100.0 milli-luminosity units per integration cycle. Each line on the diagram is for a different death rate going from 0.1 to 0.8.

input from the system's star. The effect of different natural death rates for the daisies is also illustrated as a parameter.

Conclusions

Daisyworld is the first geophysiological model. A way of viewing the world as a tightly coupled self-regulating entity and as the mathematical basis of Geophysiology. Whether right or wrong it is a source of stimulation for new thoughts and experiments. If we assume that the real world regulates to some extent like Daisyworld, then we have a source of insight for the environmental problem of biodiversity.

Biodiversity on Daisyworld is greatest when all is well with the ecosystem, but when rapid change, well within the limits of toleration, is taking place. Biodiversity is least when either the system is so stressed as to be near failure, or when it is healthy but there has been a prolonged period of steady state. With systems rich in species the relaxation time after a perturbation is very long in terms of the generation time of individual species.

Does this say anything about our present condition? We usually regard the great diversity of organisms in equatorial regions as a steady natural state. I wonder if instead we should regard this great diversity as an indication that the ecosystem or the Earth itself was healthy but has recently been perturbed. The most likely perturbation, is the sudden shift from the cool period of glaciation a mere 10,000 years ago. If this view is right then biodiversity is a symptom of change during a state of health. What seems important for sustenance is not so much biodiversity as such, but potential biodiversity, the capacity of a healthy system to respond through diversification, when the need arises. In the Amazon and other regions under threat, the damage to the biodiversity of the species may matter less than the destruction of the potential of the regional ecosystem to sustain this biodiversity.

Acknowledgments

I gratefully acknowledge helpful discussions on population and evolutionary biology with Stefan Harding, ecologist to the Dartington Trust, Totness, Devon. I wish to thank my wife Sandy Lovelock for her support and for the use of our joint income to fund this research.

References

Hassel, M. P. 1975. Density dependence in single species populations. *Journal of Animal Ecology* 44: 283-295.

Henderson-Sellers, A. and K. McGuffie, 1987. *A Climate Modelling Primer.* Chichester: Wiley.

Jascourt, S. D. and W. H. Raymond, 1991. Personal communication.

Lotka, A. 1925. *Elements of Physical Biology.* Baltimore: Williams & Wilkens.

Lovelock, J. E.. 1983. Gaia as seen through the atmosphere. In P. Westbroek and E. W. de Jong (eds.), *Biomineralisation and biological metal accumulation*, pp 15-25. Dordrecht, The Netherlands: D. Reidel.

Lovelock, J. E.. and A. J. Watson, 1982. The regulation of carbon dioxide and climate: Gaia or geochemistry, 30: 795-802.

Lovelock, J. E.. 1989. Geophysiology: the science of Gaia. *Reviews of Geophysics* 27: 215-222.

Lovelock, J. E.. 1989. *The ages of Gaia.* New York: Norton.

Maddock, L. 1991. Environmental feedback and population models. *Tellus* 43B: 331-337

May, R. M. 1974. Biological populations with non-overlapping generations: stable points, stable cycles, and chaos. *Science* 186: 645-647.

May, R. M. 1981. *Theoretical Ecology*. Oxford, U.K: Blackwell Scientific Publications.

Nicolis, G. 1991. This symposium.

Watson, A. J. and Lovelock, J. E.. Biological homeostasis of the global environment: the parable of Daisyworld. *Tellus* 35B: 284-289.

Zeng, X, R. A. Pielke, and R. Eykholt, 1990. Chaos in Daisyworld. *Tellus* 42B: 309-318.

8. Past efforts and future prospects towards understanding how many species there are

Robert M. May

Introduction

Efforts to conserve biological diversity, both in local areas and more globally, must be based on facts. What species are present in a given place? How do they respond to specific kinds of environmental disturbance and change? How do we weigh the importance of one species or community against another? What are the current global patterns of species extinction, and how -- if at all - should they be compared with the past spasms of extinction that can be seen in the fossil record?

For ill-understood reasons that lie deep in past intellectual fashion, we have made surprisingly little progress toward answering these questions. The date of the standard edition of Linnaeus' work, which may be taken as beginning the simple factual task of recording the diversity of life on Earth, is 1758. This is a full century after Newton had already given us an analytic and predictive understanding of gravitational laws, based on centuries of recorded information about planetary motions and star catalogues. The legacy symbolised by this lag between Newton and Linnaeus is still with us, and today we know more about (and spend vastly more on) the taxonomy and sytematics of stars than about the taxonomy and systematics of the organisms on our planet. We have a better estimate of the number of atoms in the universe -- an unimaginable abstraction -- than of the number of species of plants and animals currently sharing the Earth with us.

In this paper, I first sketch past patterns in recording plant and animal species, ending with an estimate of the number recorded to date. I then survey different ways of estimating what the total number of species on Earth may be; these estimates range from 3 million to 30 million or more. Next I summarise some of the very limited information about current and likely future rates of extinction. The paper concludes by noting some patterns and trends within the global "labour force" of taxonomists and sytematists, and makes some comments about the kind of *List of Recorded Species* that I would like to see given higher priority. In essentials, this paper is a condensation of material presented in more detail elsewhere (May 1988, 1990a; Gaston and May 1991).

Past Patterns in Recording Species

The 1758 edition of Linnaeus' work records around 9,000 species of plants and animals. Table 1 summarises estimates of the numbers of animal species recorded since then, for different groups, up to 1970 (after Simon, 1983). Table 1 also gives the time it took to record the second half of the total number recorded in each group, and the rough dates when new species in each group were being discovered at the fastest rate (I would have liked to have a more recent date than 1970, but I am unaware of any study more recent than Simons'). This Table confirms impressions of the different attention paid to different groups: half of all known bird species were already on "twitchers" lists within a century of Linnaeus starting it all, whereas half the arachnid and crustacean species known in 1970 were recorded in the preceding 10 years. For many, if not most, groups the peak years of discovery were in Victorian times.

Table 1. Taxonomic activity from 1758 to 1970, for different animal groups as revealed in patterns of recording new species (after Simon 1983).

Animal Group	Estimated number of species recorded up to 1970	Length of time, prior to 1970, to record the second half of the total in the previous column (years)	Period of maximum rate of discovery of new species
Protozoa	32 000	21	1897-1911
"vermes"	41 000	28	1859-1929
Arhropoda (excl. insects)	96 000	10	1959-1929
Arthropoda (insects only)	790 000	55	1859-1929
Coelenterata	9 600	58	1899-1928
Mollusca	45 000	71	1887-1899
Echinodermata	6 000	63	1859-1911
Tunicata	1 600	68	1900-1911
Chordata			
Pisces	21 000	62	1887-1929
Amphibia	2 500	60	1930-1970
Reptilia	6 300	79	1859-1929
Aves	8 600	125	1859-1882
Mammalia	4 500	118	1859-1898

The mammals and birds are, of course, very well known by now. As reviewed in detail by Diamond (1985), only 134 bird species have been added to the total of just over 9,000 since 1934, representing a rate of discovery of around 3 species per year since 1940 (most of them small, brown tropical birds). Rates of discovery are somewhat higher for mammals, with 134 of the current total of 1050 genera added since 1900, at a rate of about 1 genus per year since 1940 (most of them tropical bats, rodents or shrews, or small marsupials). As will be further emphasised in the next section, however, the story is altogether different for insects and other smaller creatures.

Table 2. The number of species (to within an order of magnitude) in the different animal phyla, classified according to habitat of adult animals. Most phyla are predominantly marine and benthic, some exclusively so. The numbers 1 through 5 indicate the approximate number of recorded living species: 1 means 1 to 10^2; 2 means 10^2 to 10^3; 3 means 10^3 to 10^4; 4 means 10^4 to 10^5; and 5 means 10^5 or more. Abbreviations: **B** = benthic; **P** = pelagic; **M** = moist; **X** = xeric; **Ec** = ecto; and **En** = endo (modified from May 1988, following Pearse 1987)

Phylum Subphylum	Habitat							
	Marine		Freshwater		Terrestrial		Symbiotic	
	B	P	B	P	M	X	Ec	En
Porifera	3		1					1
Placozoa	1							
Orthonectida								1
Dicyemida								1
Cnidaria	3	2	1	1			1	
Ctenophora	1	1						
Platyhelminthes	3	1	3		2		1	4
Gnathostomulida	2							
Nemertea	2	1	1		1		1	
Nematoda	3	1	3	1	3	1	3	3
Nematomorpha								2
Acanthocephala								2
Rotifera	1	1	2	2	1		1	1
Gastrotricha	2		2					
Kinorhyncha	2							
Loricifera	1							
Tardigrada	1		2		1			
Priapula	1							
Mollusca	4	2	3		4	1	2	1
Kampozoa	1		1				1	
Pogonophora	2							
Sipuncula	2				1			
Echiura	2							
Annelida	4	1	2		3		2	
Onychophora					1			
Arthropoda								
Crustacea	4	3	3	2	2		2	2
Chelicerata	2	1	2	2	4	3	2	1
Uniramia	1	1	3	2	5	3	2	2
Chaetognatha	1	1						
Phoronida	1							
Brachiopoda	2							
Bryozoa	3		1					
Echinodermata	3	1						
Hemichordata	1							
Chordata								
Urochordata	3	1						
Cephalochordata	1							
Vertebrata	3	3	2	3	3	3	1	1

Publication rates provide another measure of the differential attention paid to different groups (see Table 3 in May 1988). Of papers listed in the *Zoological Record* over the past few years, mammals and birds average about 1 paper per species per year, reptiles, fish and amphibians about 0.5 papers per species per year, whereas insects and other groups average from 0.1 to 0.01 or fewer papers per species per year.

Even within a given class, or order, different families can show different patterns in the attention they have received. Thus, looking at different families of insects, Strong *et al.* (1984) show that recorded species of whiteflies (*Aleyrodidae: Hemiptera*) and phytophagous thrips (*Thripidae: Thysanoptera)* increased dramatically this century, with a peak around 1920-40; weevils (*Curculionidae: Coleoptera*) show a gradual rise since Linnaean times, again peaking around 1920-40; whereas the papilionoid and danaid butterflies (*Papilionodea* and *Danainae: Lepidoptera*) are more like birds, with broad peaks in recording rates in the second half of the 19th Century.

Vascular plants, although less thoroughly studied than birds and mammals, in general have done much better than most animals. It is thought that most vascular plant species have been named and recorded, and consensual estimates put the likely total around 270 thousand (although figures as high as 400 thousand have been published, and, as we shall see in the next section, the actual number could be even higher).

So, how many plant and animal species have been named and recorded to date? No one knows! The records are mainly on file cards, scattered and poorly coordinated among institutions. Three frequently-cited estimates are 1.5 million (Grant 1973), 1.4 million, (Southwood 1978), and the more current and likely 1.8 million (Stork 1988). In my view, nothing so dramatises our societies' neglect of this central area of science as our lack of a central, computerised data base, and our consequent ignorance -- to within 10% -- of how many species have been recorded.

Table 2 ends this section by conveying a sense of the great variation in numbers of species found in different phyla, and in different environmental settings. The numbers are given only to within an order of magnitude, thus helping make the patterns clear (but also underlining the lack of precise numbers for many groups). The Table also shows that diversity has many meanings. If we speak of total numbers of species, then to a good approximation everything is a terrestrial insect. But many phyla are found in the sea: Ray (1985) has observed that, although the sea contains only 20% of all animal species, it contains systematically higher proportions of higher taxonomic units, culminating in 90% or more of all classes or phyla.

Species Totals: Direct And Empirical Estimates

How many species may there be in total? This section and the next summarise different approaches to this question, giving answers that range over a full order of magnitude, from 3 million to 30 million or more.

(a) **Projecting past trends.** One straightforward approach is to extrapolate past trends in recording new species. Given the uncertainties in the numbers actually recorded, such projections are subject to wide fluctuations, depending on the views taken of trends in particular groups and the statistical procedures used. Simon (1983, Table 31) has surveyed several such estimates, which range from concluding that essentially all species have already been named, to Simon's own estimate of around 6-7 million animal species in total. It must be emphasised that all trend-projecting estimates pre-date the dramatic studies of tropical insects discussed below.

Table 3. A very rough assessment of how ecological and taxonomic research effort is partitioned among Wallace's *Biogeographical Realms*, as indicated by the country of origin of the researchers

		Fraction of the world's ecologists[1],% [N = 16 000]	Fraction of the world's insect taxonomists[2], % [N = 1 200]
Palearctic	[Europe, Siberia]	35.3	53.5
Neartic	[North America]	43.0	25.0
	Subtotal	78.3	78.5
Oriental	[Near East to Far East and Malaysia,etc.]	12.0	7.3[3]
Ethiopian	[Subsaharan Africa]	2.3	3.8[4]
Neotropic	[South America]	1.7	3.5
Australasian	[includes N.Z., Papua New Guinea, Pacific]	5.7	5.9
	Total	100.0	100.0

[1] Countries of origin of authors of papers listed under "ecology" in Biological Abstracts 1982-83 (from Golley 1984)
[2] A very rough and biased estimate, derived from analysis of the countries of origin of researchers who borrowed from the insect collections at the British Museum (Natural History) between January 1986 and June 1991. The flaws in such an estimate are obvious, but it is nevertheless indicative (from Gaston and May 1991).
[3] Of the 7.3%, 3.6% comes from Japan, at the edge of the Oriental Realm.
[4] Of this 3.8%, most (2.9%) comes from South Africa.

(b) **An intuitive estimate.** A cruder, but not necessarily less reliable, estimate has been given by Raven (1985) and others, based on qualitative biological reasoning. This estimate of 3-5 million species rests on two observations. First, among well-studied groups such as birds and mammals there are roughly two tropical species for each temperate or boreal species. Second, the majority of all species are insects, for which temperate and boreal faunas are much better-known than tropical ones; overall, approximately two-thirds of all named species are found outside the tropics. Thus if the ratio of numbers of tropical to temperate and boreal species is the same for insects as for mammals and birds, we may expect there to be something like two yet-unnamed species of tropical insects for every one named temperate or boreal species. This carries us from the recorded total of 1.4-1.8 million species to the crude estimate that the grand total may be around 3-5 million.

(c) **Fraction of new species in previously-unstudied regions.** A very direct way of estimating species totals -- especially tropical insect totals -- is thoroughly to sample a group in some region that is relatively unstudied, and determine what fraction have previously been described. This is not easily done, because it is an enormous task completely to sample and then classify any group of tropical insects. One must, furthermore, worry whether the site of group is representative of more general patterns.

Hodkinson and Casson (1991) have made such a detailed study of the bug (*Hemiptera*) fauna of a moderately large and topographically diverse region of tropical rainforest in Sulawesi, in Indonesia. They found a total of 1690 species of terrestrial bugs, of which 63% were previously unknown. This led them, via two different lines of argument, to conclude that there are globally some 2-3 million species of insects (for a simplified account of their analysis, see May 1990a). When other taxa are added, Hodkinson and Casson's estimate accords roughly with the global total of 3-5 million in (b) above.

(d) **Tropical canopy faunas.** The studies by Erwin (1982, 1983) and others of beetles in the canopies of tropical trees, in Latin America and Malaysia, have led to much higher estimates of global species totals. Erwin's studies have found more than 1,100 different species of beetles in the canopies of *Luehea seemannii* trees in Panama, and comparable numbers in similar trees in other tropical places. These samples have not yet been "keyed out" -- a vast labour -- to discover how many of the species are new, so the method just described in (c) cannot be applied.

Instead, Erwin uses the following chain of argument and observation. First, he must guess how many of the 1,100 + beetle species are effectively specialised to *L. seemannii*, as distinct from being found generally on many trees. His estimate is around 160 canopy beetle species effectively specialised to a tropical tree species. This estimate is dominated by the guess that around 20% of the herbivorous beetles are thus specialised; my view is that tropical beetles as a group are likely to be significantly less specialised than temperate ones, and that Erwin's figure of 20% could easily be a factor 10 too high (for a more analytic discussion of what is meant by "effectively specialised", and of pertinent data, see May 1990a). Second, given 160 canopy beetle species per tree species, how many canopy insect species? For known insects generally, 40% are beetles; if this proportion applies in tropical canopies, we have 400 canopy insect species per tree species. Third, what about the rest of the tree? Erwin estimates 1 species elsewhere for each 2 species in the canopy, for a total of 600 insect species per tropical tree. This estimate could be conservative; ratios of 1:1 or even 2:1, rather than Erwin's 1:2, could be justified (Stork 1988; May 1990a). Fourth, Erwin cites the conventional estimate of around 50,000 species of tropical trees, which, when multiplied by 600 effectively specialised insect species per tree species, gives a global total of 30 million insects. If, on the other hand, we assumed only 2-3% of the beetle species were effectively specialised, but took a more generous view of the ratio of non-canopy to canopy species, we would be back into the range of 3-6 million insects.

It is easy to cavil at each link in Erwin's chain of argument. But I believe this work is important in providing a new and focused approach to the problem of estimating how many species there are. It does not so much answer the question as define an agenda of research.

e) **Other "projected-ratio" estimates.** Arguing that fungi may be thought of as the insects of the plant world, Hawksworth (1991 and this volume) has offered an upward revision of the total number of fungal species that is as dramatic as Erwin's for insects. Hawksworth's estimate is also similar to Erwin's in that it begins with a factual but local observation, and proceeds by a chain of plausible inferences towards a global estimate. Specifically, Hawksworth notes that currently some 69 thousand species of fungi have been recorded. In Britain and other well-studied Northern European regions, the ratio of vascular plant species to fungal species is around 1:6 or so. If this ratio applied generally to the estimated total of 270 thousand species of vascular plants, we arrive at something like 1.6 million species of fungi, or more than 20 times the number now known. While we must, as always, worry about applying temperate-zone ratios to tropical communities (May 1991), Hawksworth's study forces us to think more analytically about the ecological, as well as the taxonomic, assumptions underlying earlier assessments.

Another question highlighted by the work of Erwin, Hawksworth, and others is the extent to which very different patterns may pertain to different taxonomic groups, or in different settings. As discussed more fully elsewhere (May 1990a), we already see hints of this in tropical canopy faunas. Tobin's (private communication) preliminary analysis of his tropical canopy data from Manu National Park in Amazonian Peru suggests ants constitute 70% of the individuals, and beetles less than 10%, but that there are many more beetle species than ant species. In Stork's (1991) canopy samples from Bornean trees, ants were the most abundant group (4,489 of 23,874 individuals, or 19%), but they contribute relatively few species (99 of 2,800, or 3.5%); one beetle family contributes 1,455 individuals (6% of the total) but 739 species (26% of the species total). I think that patterns of "effective specialisation", which are so crucial for estimates like Erwin's, may also vary greatly among groups. In short, patterns of diversity differ significantly from group to group, which can confound any global extrapolation from a particular group or a particular place.

(f) **Polling the experts**. Gaston (1991a) has recently put together an order-by-order estimate of the global total of insect species -- the number which currently dominates any estimate of the total numbers of all species -- by canvassing the opinions of experts in each major group. Dr. Johnson may have dismissed such an assessment as a "compendium of gossip", but I think that Gaston's estimate of fewer than 10 million insect species, and probably around 5 million, is as good as we have at present.

Species Totals: Indirect and Theoretical Estimates

Our ultimate aim in recording biological diversity is not just to stick the stamps in the album, but rather to have factual foundation for asking fundamental questions

about evolution and ecology. More specifically, taxonomic lists serve as point of departure for studying the structure of food webs, patterns in the relative abundance of species,
patterns in the number of species or number of individuals in different categories of
physical size, and general trends in the commonness and rarity of organisms (for a more
concrete review, see May 1988). Interestingly, some of these studies suggest broad rules
that enable us to make independent and indirect estimates of global species totals.

(a) **Species-size relations.** One such estimate comes from observed patterns in
the numbers of terrestrial animal species in different body-size categories (May 1978,
1988). Very roughly, as one goes from animals whose characteristic linear dimension is
a few metres down to those of around 1cm (a range spanning many orders-of-magnitude
in body weight), there is an approximate empirical rule which says that for each tenfold
reduction in length (1,000 fold reduction in body weight) there are 100 times the number
of species.

This empirical relation begins to break down at body sizes below 1cm in
characteristic length. As the relation itself is not understood, this break-down may mean
nothing. But the break-down may plausibly be ascribed to our incomplete record of
smaller terrestrial animals, most of which may be unrecorded tropical insects. If the
observed pattern is arbitrarily extrapolated down to animals of characteristic length
around 0.2mm, we arrive at an estimated global total of around 10 million species of
terrestrial animals (May 1988). This frankly phenomenological estimate would be more
interesting if we had a better understanding of the physiological, ecological or evolutionary factors generating species-size distributions.

(b) **Plant-insect ratios.** If there were general patterns in the structure of
different kinds of food webs (Pimm *et al.* 1991), then it would be sufficient to know how
many plant species there are. Total species diversity would follow, computed from food
web structures. We are a long way from this grail.

Gaston (1991b) has, however, extended earlier work (Strong *et al.* 1984) by
collecting such evidence as is available for the average numbers of insect species per
plant species in particular communities. Such data comes from communities of various
sizes, in a variety of locations (from boreal to tropical). Very roughly, and with many
exceptions, there appear to be around 10 insect species per plant species. Given the
fairly secure estimate of about 270 thousand plant species, Gaston thus arrives at an
extremely crude estimate of about 3 million insect species.

(c) **Other considerations in counting species.** Many issues have been swept
under the rug in the foregoing discussion. For one thing, just what do we mean by a
species? And when we talk about microbial species, do we really mean the same thing
as when we talk about a vertebrate species (May 1988, 1990a)?

Setting aside these deeper questions, it is likely that some less fashionable
groups may be vastly more diverse than is conventionally acknowledged. Mites could
easily turn out to be as diverse as beetles; being smaller, they have received less attention.
It is quite likely that each plant and animal species has at least one parasitic nematode
and/or protozoan species that is specific to it; this could double or treble the global total
at a stroke. Recent studies show the diversity in natural populations of microbial

organisms is far greater than that found in conventional laboratory cultures (Ward *et al.* 1990; Giovannoni *et al.* 1990). We have not yet begun to address the questions that such studies of naturally occurring microbial diversity raise, for example, for the release of genetically engineered organisms.

Extinction Rates

Estimates of current extinction rates, essentially all associated with human activities, are uncertain even for comparatively well-known groups. Diamond (1989) has noted that the *Red Data Book* of the International Council for Bird Preservation (ICBP) lists 88 bird species as having become extinct since 1600, and 283 as extant but endangered, which constitutes 1 and 3% of bird species, respectively. The 88 extinctions have been at a exponentially growing rate, with about half being in this century. The lengthy procedures required for inclusion in the *Red Data Book*, however, have the consequence that the book was seriously out-of-date by the time it was published. Moreover, it tended to exclude species from less studied regions, which often are tropical regions under great pressure. A new ICPB study produced in 1988 listed 1029 bird species at risk of global extinction, and a further 637 species considered significantly threatened. This adds up to around 20% of all bird species either extinct or at risk.

Diamond (1989) goes beyond this to observe that in the areas he knows best -- New Guinea, Melanesia and Indonesia -- some of the species listed as "threatened" are surely extinct, because they occur on small islands and have not been seen for 50 years by residents of those islands who knew the species well. For instance, in the Solomon Islands the *Red Data Book* lists only 1 extinct bird species, but Diamond found no records of sightings of 12 of the Solomons' 164 bird species since 1953; most of these 12 are surely extinct. For these reasons, the ICBP is shifting its emphasis from "Red Books" of species proven to be extinct, toward "Green Lists" of species proven to be extant and secure. Projecting from his knowledge and experience, Diamond suggests that less than half of the world's roughly 9,000 species of birds would qualify for inclusion in such a catalogue of security.

Similar figures pertain to other relatively well-studied groups. For example, a recent 4-year search for the 266 species of freshwater fishes recorded from the Malay Peninsula found only 122 -- fewer than half -- of the species. There is fairly general agreement that these figures are representative (and even conservative, bearing in mind the much greater tendency to extreme endemism for most tropical insects).

More generally, roughly 1-2% of the total area of tropical forest is currently being cleared each year. This is an area roughly the size of the United Kingdom. These clearance rates are driven largely by inexorable population pressures, and they are likely to continue. The result will be a continuing disappearance of tropical forests, with consequent extinction of plant and animal species at rates we can only guess at (Myers 1989). It is reasonable to suggest that something like half of all terrestrial species are likely to become extinct over the next 50 years, if current trends persist.

Against this background, we need to go beyond knowing how many species there are, and to use this knowledge to optimise conservation efforts (either in planning

National Parks and other protected regions, or in devising strategies that reconcile sustainable exploitation of resources with preservation of an appreciable fraction of the original fauna).

Such conservation efforts will pose increasingly difficult choices. One among many questions is the extent to which all species should be treated as equal. Current concerns are largely governed by sentiment for creatures with fur or feathers. Concern diminishes as we go along a continuum from primates, to other vertebrates, to invertebrates, and it changes sign as we move down to bacteria and viruses; no one mourned the passing of the last smallpox virus. While it is presumably possible to construct an ethical framework for these feelings, it is more difficult to distinguish among otherwise similar organisms whose differences lie in their evolutionary lineages. Should all reptile species, for instance, be valued equally? Or should we take the view that, for conservation purposes, a species not closely related to other living species is more important than one with many widely distributed congeners? And if the answer to this latter questions is yes, how do we quantify the relative importance of different species?

Vane-Wright *et al.* (1991) have made a beginning on this task, showing how taxonomy and systematics can build from species lists toward assessment of the relative distinctness of different species and, ultimately, communities. At one extreme we could of course regard all species as equally important. At the opposite extreme, we could take a phylogenetic or cladistic tree, which represents the hierarchical relations among the constituent species, and measure taxonomic distinctness by weighting each group equally with respect to the summed weights of their terminal taxa. This scheme has the merit of recognising taxonomic distinctness, but it has the fault that taxonomic rank overwhelms species numbers: on this basis, the two Tuatara species in New Zealand would be seen as equal to all 6,000 other reptile species taken together. Vane-Wright *et al.* propose an intermediate scheme, which quantifies the amount of information contained in a given hierarchical classification. Their method gives answers that depend on the topology of the hierarchy (even for a fixed number of terminal taxa), and that recognise the singularity of the Tuatara without amplifying it out of all proportion (May 1990b).

What We Should be Doing, and Why

(a) **The labour force.** How big is the labour force of taxonomists and sytematists? And what regional patterns does it show? These are central questions that have received little attention.

It is easy to see that current numbers of people, and methods of working, are inadequate for the task at hand. For example, at current labour levels and with current technology and scholarly canons, Prance and Campbell (1990) estimate it would take about 380 years to complete the *Flora Neotropica*, and 950 years to complete the inventory of fungi for the same region (the latter is a pre-Hawksworth estimate). Anecdotal impressions and some data (Edwards *et al.* 1985; Parnell 1991) suggest the age-distribution of taxonomists in North America and Europe is heavily skewed to older ages, compared with other scientific disciplines. Parnell's analysis of data from the *Index Herbariorum* shows that only 13% of tropical plant taxonomists in European universities are less than 40 years old.

The distribution of taxonomists among countries is also relevant. Table 3 gives a very rough impression of how practising insect taxonomists in particular, and ecologists more generally, are distributed among Wallace's conventional "biogeographical realms". The numbers in Table 3 admittedly have many biases and other shortcomings, but they are nonetheless indicative. Both for insect taxonomy and for ecology, around 80% of the world's work force is based in North America or Europe. Only 4-7% is based where most of the biological diversity is, in Subsaharan Africa and Latin America. The relatively encouraging 7% ecologists and 12% taxonomists in Wallace's Oriental realm are mainly based in Japan, not in the Oriental tropics. These and other crude estimates are discussed more fully elsewhere (Gaston and May 1992).

Parnell's (1991) study of tropical plant taxonomists in universities shows a somewhat happier picture, with the global total increasing from around 200 or so in the early 1960s to something like 500 or more in 1981. About half this 1981 total is based in tropical universities (whose establishment or expansion over the past three decades accounts for most of the growth in numbers of tropical plant taxonomists). Encouraging as this growth in tropical plant taxonomists may look, the numbers clearly remain grossly inadequate for the task.

(b) **A List of Recorded Species**. I believe there is pressing need for a List of Recorded Species, computerised and readily available. Planning for a catalogue of all plant species, with some of the desired features, is under way, well ahead of any such synoptic plans for all animal species. But this plant catalogue, as planned by taxonomists, is simultaneously more and less than is needed for many practical and scholarly purposes. I envision a List that should be, on the one hand "quick and dirty" (at least in the first instance) with respect to formal aspects of taxonomy. But it should, on the other hand, include ecological information about the species: estimated geographical range; life history and habitat; physical size of the organism; abundance/rarity (which in many cases will be a very rough guess), and so on. Such a list of all recorded species could be an increasingly indispensable tool both for fundamental research directed toward understanding the factors that create and maintain diversity, and for practical studies directed toward assigning conservation priorities in an analytic way (see also Bisby and Hawksworth 1991).

The geographical mismatch between the home bases of the labour force and the regions where biological diversity is concentrated, as documented in Table 3, implies special responsibilities for patterns of research expenditure in developed countries. A distinctive feature of taxonomic research which is not often recognised is that, being collection-based, it takes an unusually long time to build a centre of real strength. In most other scientific disciplines, the commitment of resources can produce national strength within roughly the time it takes to train a new generation. But the "inertial effects" inherent in a collection-based discipline put long time-lags into the system. This point emerges explicitly in Parnell's (1991) analysis of tropical plant taxonomy: by appropriate initiatives, tropical universities have rapidly increased their numbers of plant taxonomists; but the total number of specimens held in herbaria in tropical universities remains relatively small. Some of these "inertial" problems can be bypassed with more imaginative approaches (including "barefoot taxonomy"), but others require developed countries fully to accept the special responsibilities that past history has left them with.

(c) **What is special about the study of biological diversity?** There are utilitarian reasons for cataloguing and studying biological diversity. Essentially all modern medicines and other pharmaceutical products have been developed from natural products, and so we should be looking at the other shelves in the larder rather than destroying them. The triumphs of intensive agriculture have been accompanied by progressive narrowing of the genetic diversity of the plants we use. The likelihood of global changes in climate give fresh emphasis to desirability of conserving existing gene pools and exploring the possibility of utilising new plants.

It is not enough to know which plant and animal species are present, locally or globally. We also need to understand how individual populations and communities of interacting populations are affected by habitat fragmentation, by changes in their physical environment (such as long-term changes in temperature or soil moisture, resulting from global climate change), and so on. Policy choices require a better knowledge of the degree to which particular ecosystems continue to function when consistuent species are extinguished or substituted; of the relation between diversity and community stability. More generally, if we are to make reliable predictions about climate change, we need to know more about how physical and ecological systems influence each other (e.g., to what extent do tropical forests generate cloud cover or buffer CO_2 inputs; never forget that our oxygen-rich atmosphere was produced by living organisms). How, ultimately, may biological diversity influence climate?

Most important, however, I believe we need to understand the diversity of living things – how much is there, and why – for the same reasons that compel us to reach out toward understanding the origins and eventual fate of the universe, or the structure of the elementary particles that it is built from, or the sequence of molecules within the human genome that code for our own self-assembly. Unlike these other quests, understanding and conserving biological diversity is a task with a time limit. The clock ticks faster and faster as human numbers continue to grow, and each year 1–2% of the tropical forests are destroyed. Future generations will, I believe, find it incomprehensible that Linnaeus still lags so far behind Newton, and that we continue to devote so little money and effort to understanding and conserving the other forms of life with which we share this planet.

References

Bisby, F. A. and D. L. Hawksworth, 1991. What must be done to save systematics? In D. L. Hawksworth (ed.), *Improving the Stability of Names: Needs and Options*. Konigstein: Koeltz Scientific.

Diamond, J. M. 1985. How many unknown species are yet to be discovered? *Nature* 315: 358-359.

Diamond, J. M. 1989. The present, past and future of human-caused extinctions. *Phil. Trans. Roy. Soc*. B325: 469-478.

Edwards, S. R., G. M. Davis and L. I. Nevling (eds.) 1985. *The Systematics Community*. Larence, Kansas: Association of Systematic Collections.

Erwin, T. L. 1982. Tropical forests: their richness in Coleoptera and other arthropod species. *Coleopt. Bull*. 36: 74-82.

Erwin, T. L. 1983. Beetles and other insects of tropical forest canopies at Manaus, Brazil sampled by insecticidal fogging. In S. L. Sutton (ed.), *Tropical Rain Forest: Ecology and Management*, p. 59-75. Oxford: Blackwell.

Gaston, K. J. 1991a. The magnitude of global insect species richness. *Conserv. Biol*. 5: 283-296.

Gaston, K. J. 1991b. Insect species: plant species ratios and estimates of global insect species richness. MS (The Natural History Museum, London).

Gaston, K. J. and R. M. May, 1992. Patterns in the numbers and distribution of taxonomists. *Nature* 356: 281–282.

Giovannoni, S. J., T. B. Britschgi, C. L. Moyer and K. G. Field, 1990. Genetic diversity in Sargasso Sea bacterioplankton. *Nature* 245: 60-63.

Golley, F. B. 1984. Introduction. In J. H. Cooley and F. B. Golley (eds.), *Trends in Ecological Research for the 1980's*, pp.1-4. New York: Plenum Press.

Grant, V. 1973. *The Origin of Adaptations*. New York: Columbia University Press.

Hawksworth, D. L. 1991. The fungal dimension of biodiversity: magnitude, significance, and conservation. *Mycol. Res*. 95: 441-456.

Hodkinson, I. D. and D. Casson, 1991. A lesser predilection for bugs: Hemiptera (Insecta) diversity in tropical rain forests. *Biol. J. Linn. Soc*. 43: 101-109.

May, R. M. 1988. How many species are there on earth? *Science* 241: 1441-1449.

May, R. M. 1990a. How many species? *Phil. Trans. Roy. Soc*. B330: 293-304.

May, R. M. 1990b. Taxonomy as density. *Nature* 347: 129-130.

May, R. M. 1991. A fondness for fungi. *Nature* 352: 475-476.

Myers, N. 1990. Mass extinctions: what can the past tell us about the present and the future? *Paleogeog., Paleoclim. and Paleoecol*. 82: 175-185.

Parnell, J. 1991. The role of Universities in tropical plant taxonomy research. MS (Botany School, Trinity College, Dublin).

Pearse, V. 1987. *Living Invertebrates*. Oxford: Blackwell.

Pimm, S. L., J. H. Lawton and J. E. Cohen, 1991. Food web patterns and their consequences. *Nature* 350: 669-674.

Prance, G. T. and D. G. Campbell, 1990. The present state of tropical floristics. *Taxon* 37: 519-548.

Raven, P. H., 1985. Disappearing species: a global tragedy. *The Futurist* 19 (5): 8-14.

Ray, G. C. 1985. Man and the sea - the ecological challenge. *Amer. Zool.* 25: 451-468.

Simon, H. R. 1983. *Research on Publication Trends in Systematic Zoology*. Ph. D. Theseis, The City University, London.

Southwood, T. R. E. 1978. The components of diversity. In L. A. Mound and N. Waloff (eds.), *Diversity of Insect Faunas*, pp. 19-40. Oxford: Blackwell.

Stork, N. E. 1988. Insect diversity: facts, fiction and speculation. *Biol. J. Linn. Soc.* 35: 321-337.

Stork, N. E. 1991. The composition of the arthropod fauna of the Bornean lowland rain forest trees. *J. Trop. Ecol.* 7: 161-180.

Strong, D. R., J. H. Lawton and T. R. E. Southwood, 1984. *Insects on plants: community patterns and mechanisms*. Oxford: Blackwell.

Vane-Wright, R. I., C. J. Humphries and P. H. Williams, 1991. What to protect? Systematics and the agony of choice. *Biol. Conserv.* 55: 235-254.

Ward, D. M., R. Weller and M. M. Bateson, 1990. 16S RNA sequences reveal numerous uncultured microorganisms in a natural community. *Nature* 345: 63-65.

9. Biodiversity in microorganisms and its role in ecosystem function

David L. Hawksworth

Introduction

Microbiologists, with a few notable exceptions, have been slow to contribute to discussions on biodiversity. The result is that such debates rarely mention, let alone consider, microorganisms. Yet, by virtue of their diversity, numbers, and roles, microorganisms are crucial to the maintenance of ecosystems at the individual, community and global levels. The field is immense, and a special workshop has been organized under the auspices of IUBS, IUMS and SCOPE to provide an authoritative statement on these issues. This present contribution is something of a *pot pourri* designed to provide the essence of the issues to be considered. It is made without prejudice to the findings of that workshop (Hawksworth and Colwell 1992a,b), which should be consulted as complementary to this overview.

Diversity

No succinct definition of "microorganism" is possible; even a trite "organism studied by a microbiologist" is untenable as most "microbiologists" pass by algae and protozoa, and commonly even ignore fungi other than yeasts. "Microbiology" itself is an unfortunate name as the field is anything but little, and all biologists who study them are not small (Cowan 1978). It is also a regrettable term in that it immediately implies something apart from, or of lesser importance than, mainstream biology, which is clearly the reverse of what it should be. Here "microorganisms" are interpreted *either* as ones belonging to phyla many members of which either cannot be seen by the unaided eye, *or* where microscopic examination, and in many cases growth in pure culture, is essential for their identification. The size range is from particles of nucleic acids in plasmids and viruses visible only by transmission electron microscopy (TEM) and measured in nm units, to massive bracket fungi (*Rigidoporus ulmarius*) up to 4 m in circumference and seaweed fronds (*Macrocystis pyrifera*) to 60 m in length (McFarlan 1990).

For convenience, terms such as "algae", "protozoa", and "fungi" are used in their traditional non-taxonomic sense, as already independently advocated (Christensen 1990; Corliss 1991; Hawksworth 1991a). This practice masks a tremendous diversity in phylogenetic and genetic terms. For example, of the 95 phyla recognized by Margulis and Schwartz (1988) as present on Earth, only 43 come within the kingdoms Plantae and Animalia. The remaining 52 phyla all belong to the "microorganisms." Further, those authors do not consider viruses on the grounds that they are not cells nor composed of cells. However, viruses fulfil the definition of life as "entities with the properties of multiplication, variation, and heredity" (Maynard-Smith 1986).

It also has to be recognized that genetic diversity at the molecular level within many microorganism groups far exceeds that of "plants" and "animals". Indeed, this diversity can be so great within microorganisms that the scientific justification for recognizing "plants" and "animals" as distinct at the kingdom level can be called into question (Sogin *et al.* 1989). The Eubacteria and Archaebacteria are certainly more distinct from each other than are the eukaryote kingdoms, and molecular studies are greatly contributing to our understanding of phylogenetic relationships at these high ranks (Solignac *et al.* 1991; Hori this volume).

At the species level, the problem of comparability between concepts also arises. Genetic distances between bacterial species of the same genus can be considerable, and may be found to exceed those between different vascular plant families. In the fungi, the few species studied in depth genetically have been found to show substantial variation especially in nutritional and biochemical attributes, and further in incompatibility groups.

When considering the conservation of the world's genetic diversity, there is no escape from the conclusion that the bulk of that diversity is exhibited in microorganism phyla, and not in "plants" and "animals". This needs to be appropriately reflected in world conservation strategies.

Magnitude

The number of formally named and accepted microorganisms is currently around 157 000 species, the fungi comprising almost half of these. However, current conservative estimates of the probable world species totals are as high as 1.8 million, the fungi and viruses arguably being the least well-known (Table 1). Microorganisms thus vastly outnumber all other groups of organisms, with the sole exception of the insects; the final total could exceed 5 million species.

About 1 800 species of fungi and 120 species of bacteria are described as new to science each year. These numbers are limited principally by a shortage of systematists available to undertake basic descriptive work. The level of our ignorance is such that previously undescribed fungi can still be found relatively easily even in Europe. Indeed, the number known in the UK has increased by 65 % during the last fifty years to around 12 000 (Hawksworth 1991a). In studies of tropical regions and hitherto little-explored habitats, it is not unusual for 15-30 % of the fungi found in limited investigations to prove to be unknown to science. As such studies are in no way comprehensive, focussing on the common and the more visible, they do not negate current estimates of fungal species, as they would if such studies were comprehensive (May 1991; this volume). Indeed, long-term studies of tropical macromycetes yield percentages of novelties around 70% (Hawksworth 1993).

Habitats not previously investigated or with extreme environments are par-ticularly important sources of novel microorganisms. Amongst a plethora of examples are: bacteria under the salt crusts of dried lakes in the East African Rift valley (Grant and Tindall 1986); bacteria associated with deep-sea hydrothermal vents (Jannasch 1990); fungi and algae living cryptoendolithically inside Antarctic rock (Friedmann 1982); and "mollicutes" (bacteria-like microorganisms without a cell wall) inside insects where it has been suggested one genus, *Spiroplasma*, may include as many as a million species and be the largest genus on earth (Whitcomb and Hackett 1989).

Table 1. The numbers of known and conservatively estimated total microorganism species in the world.

Group	Known Species	Total Species	Percentage known (%)
algae	40 000	60 000	67
bacteria	3 000	30 000	10
fungi	69 000	1 500 000	5
protozoa	40 000	100 000	40
viruses	5 000	130 000	4
Total	**157 000**	**1 820 000**	**9**

Note: Group names are used in their colloquial and not their formal taxonomic sense. The total of estimated species for bacteria does not allow for unculturable species and "mollicutes" which could together add over 3 000 000 additional species (see text and Trüper 1992).
Sources: Corliss (1991), Di Castri & Younès (1990), and Hawksworth (1991a).

Molecular probes also reveal that hitherto undetected microbes that cannot yet be isolated into culture and characterized are widespread, for example in thermal pools in the Yellowstone National Park (Ward *et al.* 1990). The current estimates of the numbers of species of bacteria and other prokaryotes (Table 1) may thus prove to be gross underestimates (Trüper 1992).

In order to obtain independent scientific evidence to test hypotheses with regard to the numbers of species of microorganisms, it is essential to study particular sites in depth. Yet, I do not know of a single site, even in Europe, which can be considered as representatively, let alone exhaustively, inventoried for all microorganism groups.

Since 1969 I have been accumulating data on the fungi present in the 185 ha Slapton Ley Nature Reserve in south Devon in the UK. Even though some 1915 fungal species have now been recorded, including 54 new to the British Isles and 21 new to science, every trip by myself, and especially by specialists in other groups, continues to yield additional fungi -- and many of the ecological niches in this site remain little-studied. These difficulties arise because of the minute size of species, problems of detection 'and identification, seasonality, and substrate or host dependence. The true number of fungi present in this single site may well exceed 2 500 species.

The difficulties in inventorying a site for fungi are more complex in the case of bacteria, protozoa, and viruses. Even a new genus of red bacteria has been cultured from a tributary of the River Thames (Austin and Moss 1986). In order to provide firmer data, selected in-depth studies need to be carefully planned and executed. Such studies, recognizing that they will nevertheless be incomplete, need to be carried out in perhaps 10 diverse sites through different biomes in the first instance. Repeated sampling would be required over a time frame of not less that five years, and using standardized sampling and isolation procedures. Ideally the same personnel should be used to compensate for observer bias. Visits would need to be planned at comparable frequencies and seasons. The scope of any such study would be enormous, even by entomological standards, as so many of the specimens would require cultural, biochemical, and ultrastructural as well as critical light microscopical study. Many would be novel, so that formal naming would be impractical in these investigations. Also, complementary expertise in the identification of host plants and insects would be vital if meaningful data regarding host specificity were to be obtained -- something crucial to providing a firmer basis for extrapolations based on species numbers (May 1991; Hawksworth 1993).

Significance in Ecosystem Function

Microorganisms played a crucial role in the evolution and diversification of the life which now exists on Earth (Margulis and Fester 1991). They continue to be of key significance in the maintenance of those life-forms at the individual, ecosystem, and global levels.

Individual level

Notwithstanding the serial endosymbiosis that gave rise to eukaryotic cells as they are recognized today, what may to a non-microbiologist appear to be a single individual of a plant or animal species, is in reality often an intimate functional biocosm involving a variety of microorganisms. For instance: herbivorous mammals such as cows, giraffes, or reindeer could not function alone to digest the cellulosic and other materials on which they depend. They rely on the microorganisms (bacteria, fungi, and protozoa) resident in their guts (Smith and Douglas 1987). About 85 % of the world's flowering plants and trees form intimate root associations with fungi, mycorrhizas, which play a crucial role in the absorption of the nutrients which most limit their growth (Read 1991); endophytic fungi ramify through healthy plant tissues, at least in some cases producing chemicals which render the plants resistant to insect pests (Carroll 1988); about 13 500 fungi harness photosynthetic algae or cyanobacteria to form lichens, swards of which dominate the ground in boreal and arctic regions (Hawksworth and Hill 1984); and the guts of many wood-boring beetles and termites always contain microorganisms, especially anaerobic bacteria and protozoa, essential for their digestion of foodstuffs (Smith and Douglas 1987).

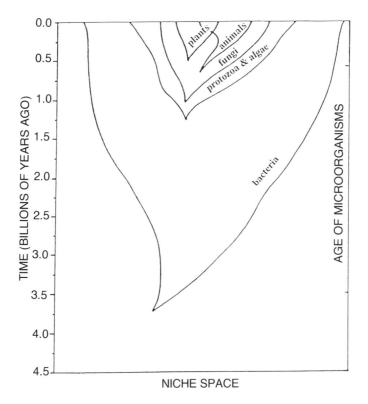

Figure 1. The nesting of groups of organisms in relation to niches occupied and evolutionary history, each group after the bacteria being dependent on earlier ones for food and mutualists. Adapted from Price (1988).

The widespread nature of such intimate, interdependent, and often coevolved associations (or "mutualisms"), amongst "plants" and "animals", and their vital role in the very existence of those life-forms, merits greater appreciation. These associations are, however, generally ignored by those concerned with the biodiversity and conservation of "macroorganisms". There is also relatively little sound information as to the degree of specificity, especially below the morphospecies level, between macroorganisms and their mutualists. The scope for informative and original experiments is immense.

Ecosystem level

However, it is the role of microorganisms in the functioning of the world's ecosystems which is of paramount importance to the maintenance of biodiversity. Food pyramids of more complex organisms rest on wide microorganism bases. The importance of different groups in food webs and the niches they are able to colonize tend to reflect evolutionary history (Price 1988; Fig. 1). This interdependence is not always appreciated. For example, numerous beetles feed on fungi; in one study in Sulawesi, of 6 236 beetle species trapped, 24 % were direct fungus feeders (Hammond 1990; Fig. 2).

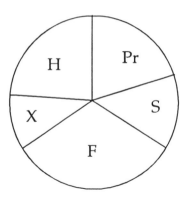

Figure 2. Feeding habits of 6 236 beetle species in a lowland forest area in N. Sulawesi. **H** = herbivorous; **Pr** = predacious; **S** = saprophagous; **F** = fungivorous; **X** = xylophagous. (Adapted from Hammond 1990.)

Food chains and webs involving microorganisms are often complex. For instance, a variety of microorganisms start to break down wood even before it is ingested and subjected to a termite's gut biota; without microorganisms the food web could not exist (Price 1988) there would not only be no termites, but the food chain onwards to dependent termite-eating mammals and birds would also be broken. Such interrelationships are more common than is generally appreciated, but are only exceptionally worked out. This lack of basic scientific knowledge makes it hazardous to forecast the effect of any microorganism loss on an ecosystem or individual. Ecologists continue to focus on the easily visible (and identifiable) in their approaches to ecosystem function, when what is required is an understanding of the microorganism dimension providing the fuel that drives the system and binding it together.

Microorganisms exhibit a vast array of biochemical roles in nature (Yanagita 1990), and execute a particularly vital role in nutrient cycling. They are responsible for the breakdown of plant and animal remains, and consequently for the recycling and/or fixing of nutrients, including carbon, nitrogen, phosphorus, and sulphur. In their absence, ecosystems would soon be swamped by their own debris. The role of nitrogen fixation is especially important in ecosystem function, and involves a considerable variety of microorganisms (Sprent and Sprent 1990); the quantification of its contribution in a variety of non-agricultural ecosystems would be of interest, including the often neglected free-living and lichen-forming cyanobacteria in tropical and subtropical humid forests.

Microorganisms that are pathogens are important in ecosystem maintenance by limiting the populations of a particular species that might otherwise dominate or disrupt the community. This can be achieved either directly or by resource competition (Price et al. 1986). Many weeds flourish in countries where they are not native as they were introduced without the pathogens reducing their vigour in their original habitats. Novel organisms with potential as biocontrol agents are frequently found only where the "weeds" are native. Fungi, bacteria, and viruses also have comparable roles in the natural control of insect pest populations, which, if allowed to increase, could affect a whole ecosystem, for example by defoliating and so eliminating dominant trees.

Microorganisms are the major biomass in soils, arguably apart from roots. The fungi predominate in amounts which regularly exceed that of the bacteria present by more than ten times (Lynch and Hobbie 1988). Microorganisms, including protozoans, are crucial to the maintenance of soil structure as well as to the cycling of nutrients. Even the formation of soil itself, through the geochemical and mechanical weathering of newly exposed rock, involves microorganisms, notably lichenized fungi and bacteria (Krumbein 1988).

Perhaps the most important function of microorganisms in ecosystems is providing mechanisms of resilience to climatic and other environmental changes. In answer to the question as to why so many microorganisms with apparently similar functions exist in successful ecosystems, Perry *et al.* (1989) eloquently argue that this renders them less vulnerable to change. If a particular tree is able to form ectomycorrhizas with a large number of fungi, as is generally the case in temperate and boreal forests, the elimination or absence of a few of those may have no significant effect on the tree and so on the ecosystem it dominates. Functional "redundancy" of microorganisms in an ecosystem should therefore be seen as a positive attribute, providing increased resistance to perturbations; a more appropriate term is required for this phenomenon.

Conversely, the loss of a single "keystone" microorganism could be dramatic for an ecosystem, for example a single fungus endomycorrhizal with a dominant forest tree, or a pathogenic fungus or virus containing the population of an otherwise potentially devastating insect pest, or a mutualistic bacterium in a large herbivore's gut.

Indications of microorganism loss are of value as an early warning of changes in an ecosystem as a result of disturbance, pollutants, or climatic factors. For example, lichenized fungi respond to gaseous pollutants (Hawksworth 1990), and mycorrhizal larger fungi are affected by acid rain (Arnolds 1991). These are particularly useful early indicators of damage before trees and crops start to be adversely affected. The potential for the use of microorganisms as bioindicators of ecosystem health, which extends to soil and water communities, is still largely untapped (Salanki 1986; Hawksworth 1992).

The scientific issues relating to the role of microorganisms in ecosystem function are rarely addressed. More analyses on the lines of that edited by Grossblatt (1989), which focused on issues of plant-microbial associations related to biocontrol, are required.

Global level

It is recognized that the present composition of the Earth's atmosphere is primarily a result of the activities of microorganisms. However, it is not always appreciated that they make a significant contribution to its maintenance. Many of the "greenhouse" gasses (i.e. methane, other methylated compounds, hydrocarbons, ammonia, carbon monoxide) are of microbial origin. In addition, microorganisms are also a carbon sink, trapping substantial amounts of carbon within their tissues, through algae (especially planktonic groups) in the sea and lakes, and on land both within soils and above it (in lichens) -- also removing carbon from the atmosphere through microbial rock weathering (see above).

The role of microorganisms in relation to global climate change merits increased attention in modelling studies from two standpoints: compensating mechanisms (Rambler *et al.* 1988), and the effects of changing pathogen ranges on crops (Parry 1990), natural ecosystems, and human diseases.

Aspects of global ecology involving microorganisms are considered more fully elsewhere in this volume (e.g. Lovelock), and are not therefore expanded on further here.

Conclusions

An appreciation of the extent of the biodiversity in microorganisms and of its importance in ecosystem function is not only of academic interest. It is fundamental to our understanding of how the ecosystems we wish to preserve function, and without which scientifically based management programmes may be defective. These ecosystems are ultimately the life-support system of *Homo sapiens.*

An understanding of the role of microorganisms in natural ecosystems also provides insights into the functioning of agroecosystems, and *vice versa*, enabling people to manipulate systems in the interests of sustainability and the maintenance of the quality of the environment. The importance and potential of such studies is only now starting to be more widely recognized by agriculturalists (Hawksworth 1991b).

Further, recognizing the current massive economic value of the products of microorganisms that are already cultured, the vastness of this currently largely untapped genetic resource merits recognition. This is especially so now that genes with desired properties can be engineered into other microorganisms to facilitate commercial production. Through creating new foodstuffs, providing alternatives to agrochemicals, and increasing the productivity of crop plants through microbial nitrogen fixation and in other ways, such exploitation may in the long-term be expected to increasingly reduce the pressure on our remaining relatively unmodified ecosystems.

Microorganisms merit more respect from "macroorganism" conservationists and ecologists. It is myopic to consider the maintenance of birds, mammals, fish, trees or flowers in natural ecosystems without according due attention to the functioning of their life-support systems. While it is rather easy to see a particular kind of truck or car is not yet extinct, it requires a much closer scrutiny and extended investigations to determine if the supply of all the crucial spares, fuels, or oils is secure, even in the short-term. Perceptions need to be broadened, and increasingly penetrating scientific questions posed and tested (Solbrig 1991), to enable soundly based conservation strategies for ecosystems to be developed -- and to safeguard Earth's extant genetic resources for future generations. We must avoid taking unconscious decisions that will lead to resourcicide, death by resource depletion. This will be expensive. US$ 20 million annually for microbial systematics and ecology was recommended to the US National Science Board (Black 1989), but I suspect a figure as high as US$ 200 million per year for a decade could be needed to adequately address the questions now, and yet to be, tabled. I suggest that such costs are not inordinately so extravagant in relation to both the importance of maintaining the earth's ecosystems, and the current world expenditure on other more questionable activities.

References

Arnolds, E. 1991. Mycologists and conservation. In D.L. Hawksworth (ed.), *Frontiers in Mycology*, pp. 243-264. Wallingford: CAB International.

Austin, D. A. and M. O. Moss, 1986. Numerical taxonomy of red-pigmented bacteria isolated from a lowland river, with the description of a new taxon, *Rugamonas rubra* gen. nov., sp. nov. *Journal of General Microbiology* 132: 1899-1909.

Black, C. C. 1989. *Loss of Biological Diversity: A global crisis requiring international solutions*. Washington, D.C.: National Science Board. 19 pp.

Carroll, G. 1988. Fungal endophytes in stems and leaves; from latent pathogen to mutualistic symbiont. *Ecology* 69: 2-9.

Christensen, T. 1990. Plants, animals, algae and fungi, four non-taxonomic group designations. *Taxon* 39: 93-94.

Corliss, J. O. 1991. Introduction to the protozoa. In F. W. Harrison and J. O. Corliss (eds), *Microscopic Anatomy of Invertebrates*, vol.1, pp. 1-12. New York: Wiley-Liss.

Cowan, S. T. 1978. *A Dictionary of Microbial Taxonomy*. Cambridge: Cambridge University Press. 285 pp.

Di Castri, F. and T. Younès, 1990. Ecosystem function of biological diversity. *Biology International, Special Issue* 22: 1-20.

Friedmann, E. I. 1982. Endolithic microorganisms in the Antarctic cold desert. *Science* 215: 1045-1053.

Grant, W. and B. Tindall, 1986. The alkaline saline environment. *Society for General Microbiology, Special Publication* 17: 25-54.

Grossblatt, N. (Ed.), 1989. *The Ecology of Plant-Associated Microorganisms*. Washington, D.C.: National Academy Press. 34 pp.

Hammond, P. M. 1990. Insect abundance and diversity in the Dumoga-bone National Park, N. Sulawesi with special reference to the beetle fauna of lowland rainforest in the Torault region. In W. J. Knight and J. D. Holloway (eds), *Insects and the Rain Forests of S.E. Asia (Wallacea)*, pp. 197-254. London: Royal Entomological Society.

Hawksworth, D. L. 1990. The long-term effects of air pollutants on lichen communities in Europe and North America. In G. M. Woodwell (ed.), *The Earth in Transition. Patterns and processes of biotic impoverishment*, pp. 45-64. Cambridge: Cambridge University Press.

Hawksworth, D. L. 1991a. The fungal dimension of biodiversity: magnitude, significance, and conservation. *Mycological Research* 95: 641-655.

Hawksworth, D. L. (Ed.) 1991b. *The Biodiversity of Microorganisms and Invertebrates: its role in Sustainable Agriculture.* Wallingford: CAB International. 302 pp.

Hawksworth, D. L. 1992. Litmus tests for ecosystem health: the potential of bio-indicators in the monitoring of biodiversity. In M. S. Swaminathan and S. Jana (eds), *Biodiversity: Implications for Food Security*, pp. 184-204. Macmillan India.

Hawksworth, D. L. 1993. The tropical fungal biota: census, pertinence, prophylaxis, and progress. In S. Isaac, J. C. Frankland, R. Watling and A. J. S. Whalley (eds), *Aspects of Tropical Mycology*, pp. 265-293. Cambridge: Cambridge University Press.

Hawksworth, D. L. and R. R. Colwell (Eds) 1992a. Biodiversity amongst microorganisms and its relevance. *Biology International* 24: 11-15.

Hawksworth, D. L. and R. R. Colwell (Eds) 1992b. Biodiversity amongst microorganisms and its relevance. *Biodiversity and Conservation* 1: 221-345.

Hawksworth, D. L. and D. J. Hill, 1984. *The Lichen-forming Fungi*. Glasgow and London: Blackie. 158 pp.

Jannasch, H. W. 1990. The microbial basis of life as deep-sea hydrothermal vents. *American Society for Microbiology News* 55: 413-426.

Krumbein, W. E. (Ed.) 1983. *Microbial Geochemistry*. Oxford: Blackwell Scientific Publications. 330 pp.

Lynch, J. M. and J. E. Hobbie (Eds) 1988. *Micro-organisms in Action: concepts and applications in microbial ecology*. Second edition. Oxford: Blackwell Scientific Publications. 300 pp.

McFarlan, D. 1990. *The Guinness Book of Records 1991*. London: Guinness Publishing. 320 pp.

Margulis, L. and R. Fester (Eds) 1991. *Symbiosis as a Source of Evolutionary Innovation*. Cambridge, Mass.: Massachusetts Institute of Technology Press. 454 pp.

Margulis, L. and K. V. Schwartz, 1988. *Five Kingdoms. An illustrated guide to the phyla of life on Earth*. 2nd edition. New York: W.H. Freeman. 376 pp.

May, R. M. 1991. A fondness for fungi. *Nature, London* 352: 475-476.

Maynard-Smith, J. 1986. *The Problems of Biology*. Oxford: Oxford University Press. 134 pp.

Parry, M. 1990. *Climate Change and World Agriculture*. London: Earthscan Publications. 157 pp.

Perry, D. A., M. P. Amaranthus, J. G. Borchers, S. L. Borchers, and R. E. Brainerd, 1989. Bootstrapping in ecosystems. *BioScience* 39: 230-237.

Price, P. W. 1988. An overview of organismal interactions in ecosystems in evolutionary and ecological time. *Agriculture, Ecosystems and Environment* 24: 369-377.

Price, P. W., M. Westoby, M. Rice, P. R. Atsatt, R. S. Fritz, J. N. Thompson, and K. Mobley, 1986. Parasite mediation in ecological interactions. *Annual Review of Ecology and Systematics* 17: 487-505.

Rambler, M.B., L. Margulis, and R. Fester (Eds) 1988. *Global Ecology. Towards a science of the biosphere*. San Diego: Academic Press. 204 pp.

Read, D. J. 1991. Mycorrhizas in ecosystems - nature's response to the "Law of the Minimum". In D. L. Hawksworth (ed.), *Frontiers in Mycology*, pp. 101-130. Wallingford: CAB International.

Salanki, J. (Ed.) 1986. *Biological Monitoring of the State of the Environment. Bioindicators*. Eynsham, Oxford: IRL Press.

Smith, D. C. and A. E. Douglas, 1987. *The Biology of Symbiosis*. London: Edward Arnold. 302 pp.

Sogin, M. L., J. H. Gunderson, H. J. Elwood, R. A. Alonso, and D. A. Peattie, 1989. Phylogenetic meaning of the kingdom concept: an unusual ribosomal RNA from *Giardia lamblia*. *Science* 243: 75-77.

Solbrig, O.T. 1991. *Biodiversity. Scientific issues and collaborative research proposals*. Paris: United Nations Educational, Scientific and Cultural Organization. 77 pp.

Solignac, M., M. Pélandakis, F. Rousset, and A. Chenuil, 1991. Ribosomal RNA phylogenies. In G.M. Hewitt, A.W.B. Johnston and J.P.W. Young (eds), *Molecular Techniques in Taxonomy*, pp. 73-85. Berlin: Springer-Verlag.

Sprent, J.I. and P. Sprent, 1990. *Nitrogen Fixing Organisms*. London: Chapman and Hall. 256 pp.

Trüper, H. G. 1992. Prokaryotes: an overview with respect to biodiversity and environmental importance. *Biodiversity and Conservation* 1: 227-236.

Ward, D.M., R. Weller, and M.M. Bateson, 1990. 16S rRNA sequences reveal numerous uncultured microorganisms in a natural community. *Nature 345: 63-65.*

Whitcomb, R.F. and K.J. Hackett, 1989. Why are there so many species of mollicutes? An essay on prokaryotic diversity. In L. Knutson and A.K. Stoner (eds), *Biotic Diversity and Germplasm Preservation, global imperatives*, pp. 205-240. Dordrecht: Kluwer Academic.

Yanagita, T. 1990. *Natural Microbial Communities. Ecological and physiological features*. Tokyo: Japan Scientific Societies Press. 485 pp.

Selected more recent publications

Caroll, G. C. and D. T. Wicklow (Eds) 1992. *The Fungal Community. Its organization and role in the ecosystem*. Second edition. New York: Dekker. 976 pp.

Guerro, R. and C. Pedrós-Alió (Eds) 1993. *Trends in Microbial Ecology*. Barcelona: Spanish Society for Microbiology.

Hawksworth, D. L. and J. M. Ritchie 1993. *Biodiversity and Biosystematic Priorities: Microorganisms and Invertebrates*. Wallingford: CAB International. 120 pp.

Isaac, S., J. C. Frankland, R. Watling, and A. J. S. Whalley (Eds) 1993. *Aspects of Tropical Mycology*. Cambridge: Cambridge University Press. 325 pp.

Jones, D. G. (Ed.) 1993. *Exploitation of Microorganisms*. London: Chapman and Hall. 488 pp.

Read, D. J., D. H. Lewis, A. H. Fitter, and I. J. Alexander (Eds) 1992. *Mycorrhizas in Ecosystems*. Wallingford: CAB International. 419 pp.

Whitton, B. A. 1992. Diversity, ecology and taxonomy of the cyanobacteria. In N. H. Mann and N. G. Carr (Eds), *Photosynthetic Prokaryotes*, pp. 1-51. New York: Plenum Press.

10. Evolutionary outline of living organisms as deduced from 5S ribosomal RNA sequences

Hiroshi Hori

Introduction

In 1866, Ernst Haeckel (1866) exploited a presumed connection between ontogeny and phylogeny, suggesting that the tree of life has two main branches: plants and animals. After that he newly recognized the single-celled forms, the protists, as the third branch. Copeland (1956) later split out a fourth main branch, a new kingdom accommodating the bacteria (Monera), and Whittaker (1959) created a fifth, for the fungi, dividing the living world into Animalia, Plantae, Fungi, Protista, and Monera (the five kingdoms theory). On the other hand, based on the primary difference of the cellular structures, Chatton (1938) divided all life into eukaryotes and prokaryotes. The five kingdom scheme today includes the eukaryote-prokaryote concept, this is the most widely accepted view of the basic organization of life based on physiological and morphological characters (Margulis 1970).

These evolutionary investigations were based on the study of complex morphologies and detailed fossil records. However, the changes of these characters are very complicated and the rate of change is variable in different groups of organisms and in different evolutionary periods. Therefore not much confidence can be given to the systems. A more useful approach to this problem is to use protein, DNA or RNA sequences, because the evolutionary change of these molecules is roughly proportional to evolutionary time (= molecular clock, Zuckerkandl and Pauling 1965; Kimura 1983). Based on this idea and using 16S rRNA sequence data, Woese *et al.* (1990) claimed that life on this planet divides into three primary groupings, commonly known as the eubacteria, the archaebacteria, and the eukaryotes, denying the previous theories, i.e., the five kingdom theory and the prokaryote-eukaryote concept.

In the present paper, we compared 550 complete nucleotide sequence of 5S ribosomal RNAs from the cytoplasm, chloroplasts and mitochondria, and constructed a phylogenetic tree to elucidate the major evolutionary process of living organisms and cellular systems, especially origins of chloroplasts and mitochondria. As a result, we confirmed that the eukaryote- prokaryote concept is correct and that the Metabacteria were derived from eubacteria but are not the most ancient organism on earth, in opposition to the archaebacteria hypothesis (Woese *et al.* 1990).

Sequence data and secondary structure

Generally, the secondary-structure model of 5S rRNAs is the same for all organisms, but there exists partial specificity in each group ; secondary structures may be classified into four types, i.e., the eukaryotic type, metabacterial type, eubacterial type (including chloroplasts), and mitochondrial type (Hori and Osawa 1987).

The first class is the eukaryotic type. Eukaryotic 5S rRNA differs from all the eubacterial 5S rRNAs by having the eukaryote-specific E-E' base- pairing and the loop (Fig. 1A, C). The second class is the metabacterial type (Fig. 1B). This has the non-conserved "eukaryotic" E-E' helix, a bulge of the D-D' region and the conserved region between D and E which corresponds to RNA polymerase III recognition site on the 5S rRNA gene in the eukaryote, supporting the view that metabacteria are more related to eukaryotes than to eubacteria. The third class of 5S rRNAs, to which all eubacterial 5S rRNAs (and chloroplast 5S rRNAs) belong, may be called the eubacterial type (Fig. 1C). They have highly conserved regions corresponding to less conserved E-E' regions in eukaryotic 5S rRNAs. Fourth is the mitochondrial type (Fig. 1D). The model shown here is a 5S rRNA from a primitive green alga, *Mamiella* sp., showing a model similar to eubacterial type.

Homology matrix

On the basis of the 5S rRNA alignment (not shown), percent similarities of all possible pairs of sequences were calculated and summarized in a homology triangle of 500 representative sequences (Fig. 2). The matrix contains major representative groups of the living organisms. This scheme does not show much detail as a dendrogram, but groups with high homologies can be recognized distinctly. For example, all animals consist of a clear dark triangle, and almost all green plants, such as land plants and green algae, form a cluster. The density values between *Chromophyta* (brown algae and diatoms, etc.) and *Basidiomycota* (mushrooms and smuts) are very high, suggesting a close relationship between them. Figure 2 clearly indicates the low homology of metabacterial 5S rRNA sequences to those of eukaryotes and eubacteria. In spite of these low homologies, metabacterial sequences reveal a slight homology to those from actinobacteria, representing their presumed evolutionary relatedness.

Tree construction

Evolutionary distance between two 5S rRNA sequences, Knuc, and its SE were calculated as previously shown (Hori and Osawa 1987; Hori, Lim and Osawa 1985). Briefly, Knuc estimates the number of base substitutions per nucleotide site that have occurred since the separation of the two sequences as follows:

$$\text{Knuc} = -(1/2) \log_e [(1 - 2P - Q)(1 - 2Q)1/2],$$

where P and Q are the fractions of nucleotide sites between two sequences showing transition- and transversion-type differences, respectively. A phylogenetic tree

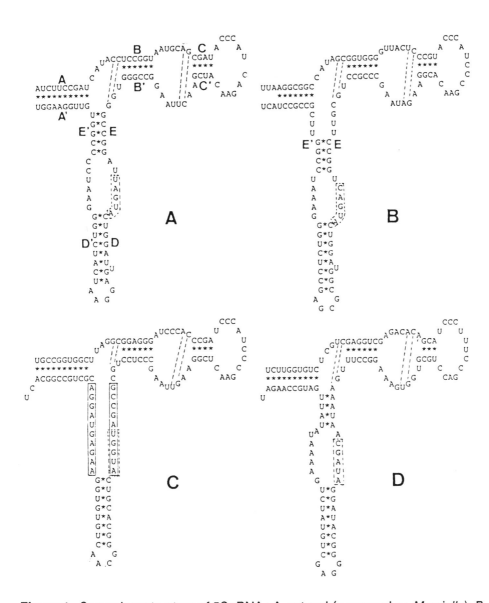

Figure 1 Secondary structure of 5S rRNA : A, cytosol (a green alga, *Mamiella*); B, metabacterial (*Halobacterium cutirubrum*); C, chloroplast (*Mamiella*); D, mitochondrial (*Mamiella*). Asterisk and dashed line indicate base-pairs and potential base-pairs, respectively. Base-paired regions are indicated with symbols A, A', B, B', etc. Regions boxed-in with a solid line in model C indicate highly conserved regions in eubacterial (and chloroplast) sequences. Those with dashed line indicate a probable recognition site on DNA with RNA polymerase III in eukaryote sequence and the corresponding conserved prokaryotic sequences (Hori and Osawa 1987).

was constructed by the simplified unweighted pair group average method (Hori and Osawa 1987).

Outline of the phylogenetic tree

A phylogenetic tree deduced from 558 5S rRNA sequences has been constructed, and an outline of it including most of the major groups of photosynthetic organisms is represented in Fig. 3. As we pointed out previously (Hori and Osawa 1987), the tree shows that eubacteria first separated from the metabacteria/eukaryotes branch. Metabacteria form a unique group that is phylogenetically closer to eukaryotes than to eubacteria. This suggests that the eukaryote ancestor is a metabacteria. Note that *Sulfolobus*, methanogens, and extreme halophiles, which collectively we called metabacteria (Hori and Osawa 1987; Osawa and Hori 1979), are called archaebacteria by Woese and his colleagues (Woese *et al*. 1990; Woese and Fox 1977). In this paper, however, we use metabacteria instead of archaebacteria, because in our view eubacteria are more ancient than to metabacteria.

While the emergence of eukaryotes from prokaryotes was the most drastic transformation ever undergone by living organisms, the origin of the first bacterial cell was even more important because it created the first true organism. The tree shows that the first cells on this planet were Gram-negative bacteria. Cavalier-Smith (1987) also proposed a similar hypothesis on the basis of the similarities between them at the molecular and cellular levels. He also suggested the possibility that the first cell was a photosynthetic gram-negative eubacterium.

Among eukaryotes, three major types of photosynthetic eukaryotes, i.e., red algae (chlorophyll a + phycobilins group), green plants (chlorophyll a + b group) and chromophyte algae (chlorophyll a + c group) - are more remotely related to one other. The divergence order of these three major plant groups indicates that a single symbiosis of photosynthetic prokaryotes with eukaryotes (= emergence of photosynthetic eukaryotes) took place before the divergence of the different kinds of algae.

The emergence point of red algae is estimated to be 1.3- 1.4 billion years ago, showing the most ancient event so far detected in the evolution of eukaryotes. The separation of red algal species took place at very early times. All green plants examined herein, belong to the same green plant branch, in accordance with the classical view (Hori *et al*. 1985).

Chromophyte algae contain the Phaeophyta (brown algae), Dinophyta (diatoms) and Chrysophyta (golden-yellow algae) because of their having chlorophylls a and c and the unique storage substance laminarin. The comparison of these 5S rRNA sequences indicates that the three groups belong to the same branch, the golden-yellow alga separating first, then the diatoms and brown algae separating from each other. Brown algae, which consists of 270 genera and 1,500 species, is one of the most morphologically diversified eukaryotic groups. However, homology percents of 5S rRNA sequences from five typical brown-algal species - which cover the representative major orders of this phylum - are very high, indicating that all brown algae separated

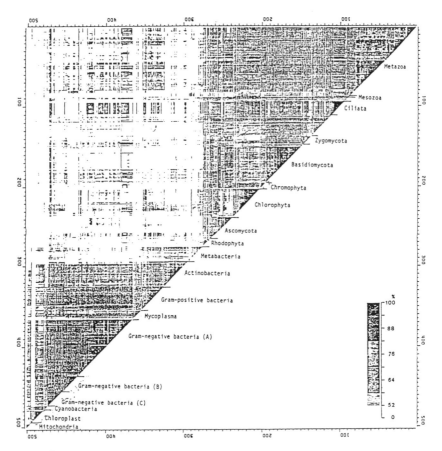

Figure 2. Density matrix of homology percents constructed from representative 500 5S rRNA sequences. Homology percent of all possible pairs of 5S rRNAs (500 x 500 matrix) was calculated in the mainframe computer, and plotted as density value by means of a program for the Versetec Printer Plotter System (upper left half of Fig. = homology triangle). The matrix indicates that the living organisms can be classified at a glance as a dense triangle. All 5S rRNA sequences from various organisms used in this study were compiled in a database IRIS, together with their taxonomic information and details of the alignment (Hori and Satow 1989).

from one another with in a very short time. The divergence point of them was 0.2 billion years ago (Hori and Osawa 1987).

The details of each branch have been reported in previous papers: for a general review see Hori and Osawa (1987); for the gram-negative eubacteria Lane *et al.* (1985); for the gram-positive eubacteria Dekio *et al.* (1984); for the eubacterial family Vibrionaceae MacDonell and Colwell (1985); for the green plants Hori *et al.* (1985); for the Ascomycota Chen *et al.* (1984); for the Basidiomycota (Walker 1984); for Protozoans Kumazaki (1983) and for Meso- and Metazoa Hori and Osawa (1987) and Ohama (1984).

Organelle evolution

It has been long believed that the autogenous origin of chloroplasts and mitochondria occurred within the eukaryotic cells by forming membrane compartments (Margulis 1970). However, molecular data accumulated by DNA sequence analysis of gene organization and composition strongly suggest that chloroplasts and mitochondria originated from cyanobacteria and purple photosynthetic bacteria, respectively (Margulis 1970; Ohyama *et al.* 1990). In the case of 5S rRNA, chloroplasts have a typical eubacterial 5S rRNA (Fig. 1C), and their phylogenetic position in the tree is very near to cyanobacteria and the cyanelle of *Cyanophora paradoxa*, suggesting the symbiotic origin of chloroplast (point A in Fig. 3). Complete DNA sequence analysis of chloroplast genomes from several species of plant cells clearly supports this view (Ohyama *et al.* 1990).

Contrary to chloroplast evolution, phylogenetic position of mitochondria in the 5S rRNA tree is still in debate. Sometimes, mitochondrial 5S rRNAs show an startling divergence point due to the influence of their unusual secondary structure (see Fig. 1d of Hori and Osawa 1987). Through comparison of some conserved regions of mitochondrial 5S rRNA from a green alga, *Mamiella*, we could obtain "more reasonable" results for the symbiosis theory (Hori *et al.* 1990).

Archaebacteria versus Metabacteria

Based on comparison of the 16S (or 18S) rRNAs, Woese and colleagues (1977, 1990) proposed that there are two fundamentally different groups of bacteria, eubacteria and archaebacteria, and that, with eukaryotes, they constitute the "three primary kingdoms" of life. Among them, the archaebacteria must be primary (Woese and Fox 1977), and not secondarily evolved from eubacteria (we call this concept the archaebacteria hypothesis). They rejected the widely accepted view that life began in the ocean, but claimed a new concept that it may have begun in enormous cloud banks surrounding a steaming, primitive Earth.

Although the existence of archaebacterial "kingdom" is generally agreed, the evolutionary relationship of the three major lineages is still subject to much debate (Van Valen and Mariorana 1981; Hori, Itoh and Osawa 1982; Lake *et al.* 1984). Since this hypothesis was originally deduced by comparing the oligonucleotide catalogs of 16S rRNA (SAB method), the actual evolutionary distances of SAB values between the three hypothetical kingdoms were reported to be nearly the same (Woese and Fox 1977), and the value itself is statistically almost meaningless (Hori, Itoh and Osawa 1982). The reason why the archaebacteria was defined as the most ancient group of bacteria yet detected was only from resemblance of their habitat presumed to exist on earth 3-4 billion years ago (Woese and Fox 1977).

A phylogenetic tree deduced from 5S rRNA shows eubacteria emerging first, and then metabacteria (= archaebacteria) and eukaryotes separating from each other (Fig. 3), contradicting the archaebacteria hypothesis deduced from 16S rRNA data.

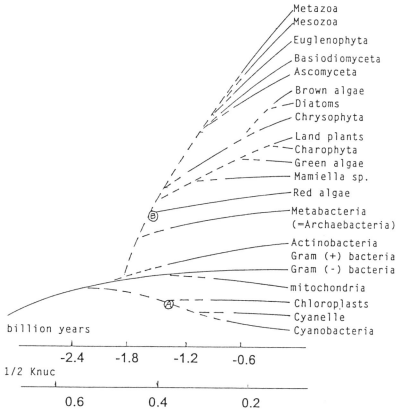

Figure 3. Simplified phylogenetic tree of major groups of organisms as deduced from 558 5S rRNA sequences. Time scale (abscissa) was first estimated from the evolutionary distance, Knuc, by assuming that the rate of nucleotide substitution in 5S rRNA is constant from eubacteria to man throughout. For convenience, time scale was converted to the year estimated on the basis of yeast-animal divergence time (1.2 billion years ago; (Hori & Osawa 1987)). As previously reported (Hori & Osawa 1987), the tree was revised by adjusting the divergence points of chloroplasts and their hosts. Thus, the rate of nucleotide substitution in eubacteria- chloroplast branch is almost two times lower than those from eukaryotes-metabacteria branch.

Metabacteria might be more closely related to eukaryotes than to eubacteria (= metabacteria hypothesis (Hori and Osawa 1987; Osawa and Hori 1979).

There are many data that support the metabacteria hypothesis. A recent advance in such studies is the work by Iwabe *et al.* (1989). Using DNA sequence data involving a few duplicated genes shared by all three primary kingdoms, these authors presented definitive evidence showing that archaebacteria are not old but new. In other words, archaebacteria are more closely related to eukaryotes than to eubacteria. These include information from elongation factors Tu and G, large subunit of DNA-dependent RNA polymerase, and alpha and beta subunits of Sulfolobus ATPase.

The most plausible explanation for the discrepancies between 16S rRNA and other molecules, such as 5S rRNA, could be that the ancestor common to the eubacteria and metabacteria/eukaryotes had the macromolecules of the eubacterial nature. After their separation, the eubacterial characters have been conserved in the eubacterial line, while the organisms belonging to the other stem gradually acquired the eukaryotic characters, and the bulk of them eventually lost these eubacterial characters and evolved into the eukaryotes. The remainder, which is the metabacteria, failed to do so, thus the present-day metabacteria, having both eukaryotic (in the case of ATPase, EF-G, Tu, and 5S rRNA) and eubacterial characters (in the case of 16S rRNA) in their macromolecules. At the emergence of eukaryotes from metabacteria, 16S rRNA produced a drastic conformational change, so that the rate of nucleotide substitution in 16S rRNA molecules varied among different groups of organisms (or different types of secondary structure). The non-linear evolution process might have increased substitution rate on this molecule.

Although metabacteria form a very unique group of bacteria, 5S rRNA tree supports the view that these bacteria still belong to the prokaryotic domain together with eubacteria. Therefore living organisms on this planet would be seen as comprising two domains: prokaryotes and eukaryotes. The prokaryotic domain contains eubacterial and metabacterial kingdoms as shown in Fig. 3. This does not support the concept of three domains, Archaea, Bacteria and Eukarya by Woese *et al.* (1990).

Acknowledgement

This work was supported by Grant-in-Aid No.62304006 from the Ministry of Education, Science and Culture of Japan.

References

Cavalier-Smith, T. 1987. The origin of eukaryote and archaebacterial cells. *Ann. New York Acad. Sci.* 503: 17-54.

Chatton, E., 1938. *Titres et Travoux Scientifiques (1906-1937) de Edouard Chatton.* Sete, France: E. Sottano.

Chen, M.W., J. Anne, G. Volckaert, E. Huysmans, A. Vandenberghe, and R. De Wachter, 1984. The nucleotide sequences of the 5S rRNAs of seven molds and a yeast and their use in studying ascomycete phylogeny. *Nucleic Acids Res.* 12: 4881-4892.

Copeland, H.F., 1956. *The classification of the lower organisms.* Palo Alto: Pacific Books.

Dekio, S., R. Yamasaki, J. Jidoi, H. Hori and S. Osawa, 1984. Secondary structure and phylogeny of *Staphylococcus* and *Micrococcus* 5S rRNA. *J. Bacteriol.* 159: 233-237.

Haekel, E., 1866. *Generelle Morphologie der Organismen.* Berlin: Reimer.

Hori, H. and S. Osawa 1987. Origin and evolution of organisms as deduced from 5S ribosomal RNA sequences. *Mol. Biol. Evol.* 4: 445-472.

Hori, H. and Y. Satow, 1989. IRIS: Integrated RNA information for systematics. In R.R Colwell (Ed.), *Biomolecular data: a resource in transition*, pp. 179-184. Oxford: Oxford Univ. Press.

Hori, H., T. Itoh and S. Osawa, 1982. The phylogenic structure of the metabacteria. *Zbl. Bakt. Hyg.*, I. Abt. Orig. C3: 18-30.

Hori, H., B. L. Lim and S. Osawa, 1985. Evolution of green plants as deduced from 5S rRNA sequences. *Proc. Natl. Acad. Sci. USA* 82: 820-823.

Hori, H., Y. Satow, I. Inoue, and M. Chihara, 1990. Archaebacteria vs Metabacteria: Phylogenetic tree of organisms indicated by comparison of 5S ribosomal RNA sequences. In S. Osawa and T. Honjo (Eds.), *Evolution of life: Fossils, Molecules, and Culture*, pp. 325-336. Heidelberg: Springer.

Iwabe, N., K. Kuma, M. Hasegawa, S. Osawa, and T. Miyata, 1989. Evolutionary relationship of archaebacteria, eubacteria and eukaryotes inferred from phylogenetic trees of duplicated genes. *Proc. Natl. Acad. Sci.* USA 86: 9355-9359.

Kimura, M., 1983. *The neutral theory of molecular evolution*. Cambridge, U.K.: Cambridge Univ. Press.

Kumazaki, T., H. Hori, and S. Osawa, 1983. Phylogeny of protozoa deduced from 5S rRNA sequences. *J. Mol. Evol.* 19: 411-419.

Lake, J.A., E. Henderson, M. Oakes, and M.W. Clark, 1984. Eocytes: A new ribosome structure indicates a kingdom with a close relationship to eukaryotes. *Proc. Natl. Acad. Sci.* USA 81 :3786-3790.

Lane, D.J., D.A. Stahl, G.J. Olsen, D.J. Heller and N.R. Pace, 1985. Phylogenetic analysis of the genera *Thiobacillus* and *Thiomicrospira* by 5S rRNA sequences. *J. Bacteriol.* 163: 75-81.

MacDonell, M.T. and R. R. Colwell, 1985. Phylogeny of *Vibrionaceae*, and recommendation for two new genera, *Listonella* and *Shewanella*. *Syst. Appl. Microbiol.* 6: 171-182.

Margulis, L., 1970. *Origin of eukaryotic cells*. New Haven: Yale Univ.Press.

Ohama, T., T. Kumazaki, H. Hori and S. Osawa, 1984. Evolution of multicellular animals as deduced from 5S rRNA sequences: a possible early emergence of the Mesozoa. *Nucleic Acids Res.* 12: 5101-5108.

Ohyama, K., Y. Ogura, K. Oda, K. Yamato, E. Ohta, Y. Nakamura, M. Takemura, N. Nasato, K. Akashi, T. Kanegae and Y. Yamada, 1990. Evolution of organellar genomes. In: S. Osawa and T. Honjo (Eds.), *Evolution of life: Fossils, Molecules, and Culture*, pp. 187-200. Heidelberg: Springer.

Osawa, H. and H. Hori, 1979. Molecular evolution of ribosomal components. In G. Chambliss, G.R. Craven, J. Davies, K. Davis, L. Kahan, and M. Nomura (Eds.), *Ribosomes, Structure, Function, and Genetics*, pp. 333-385. Baltimore: Univ. Park Press.

Van Valen, L. M. and V. C. Mariorana, 1981. The archaebacteria and eukaryotic origins. *Nature* 287: 248-250.

Walker, W. F. 1984. 5S ribosomal RNA sequences from *Atractiellales*, and Basidiomycetous yeasts and Fungi Imperfecti. *Syst. Appl. Microbiol.* 5: 352-359.

Wittaker, R. H., 1959. On the broad classification of organisms. *Quart. Rev. Biol.* 34: 210-226.

Woese, C. R. and G. E. Fox, 1977. Phylogenetic structure of prokaryotic domain: the primary kingdoms. *Proc. Natl. Acad. Sci. USA* 74: 5088-5090.

Woese, C. R., O. Kandler and M. L. Wheelis, 1990. Towards a natural system of organisms: Proposals for the domains *Archaea*, *Bacteria* and *Eukarya*. *Proc. Natl. Acad. Sci. USA* 87 :4576-4579.

Zuckerkandl, E. and L. Pauling, 1965. Evolutionary divergence and convergence in proteins. In: V. Bryson and H.J. Vogel (Eds.), *Evolving genes and proteins*, pp. 97-166. New York: Academic Press.

11. The role of biodiversity in marine ecosystems

Pierre Lasserre

Introduction

Oceans cover 70% of the earth's surface, they are all connected between them. They play a fundamental role in the regulation of the global environment through physical and biogeochemical processes that largely determine climate changes. The coastal areas include the Exclusive Economic Zones (EEZ) of the maritime nations and they produce 80% of the marine living resources. By year 2000, 60% of the world population will be living or depending on the coastal zone.

Until very recently, it was considered that man could not change the seas and their biological diversity, that exploitation of living resources could not possibly alter fish stocks, and that the oceans could absorb all our wastes. These views are no longer tenable. In the oceans, changes are taking place similar to those on land, and much more rapidly. In the coastal zone, significant consequences can be expected of the green house effect, and while human activities continue to extend beyond the shelf edge, even those deep-sea reservoirs of biological diversity may be at risk.

We are only now beginning to describe marine biodiversity and to appreciate the implications of its different facets, at the gene flow level, species and higher taxonomic levels, and ecological levels. The processes maintaining biological diversity in marine ecosystems are, at first sight, similar to those seen in terrestrial ecosystems: e.g. habitat fragmentation (patch dynamics). There are major differences, however, largely due to the dispersive nature of marine larvae and the wide distribution of organisms and habitats. We are far from being able to understand the importance of biodiversity to the functioning of ecosystems and their processes. Therefore, a scientific program to help understand marine biodiversity and its ecological significance is urgently required (Grassle *et al.* 1991; Ray and Grassle 1991; Solbrig 1991). Large-scale experiments are needed on whole-system responses and on carefully selected hot spots under tropical, temperate and polar latitudes, from the near shorelines to deep oceans (Ray 1991; Lasserre 1991). These experiments could have far reaching implications for the conservation and the management of oceanic ecosystems and the way in which they are eventually exploited.

In this presentation of the role of biodiversity in marine ecosystems, I will focus and restrict myself on some key questions and examples showing some connection exiting between marine biodiversity, ecosystem function and global change.

Table 1. Number of species and number of families in each phylum from 233 samples, representing a total area of 21 m^2 taken between 1500 m and 2500 m depth on the continental slope of New Jersey (from Grassle *et al.* 1990).

Phylum	Number of species	Number of families
Cnidaria	19	10
Nemertea	22	1
Priapulida	2	1
Annelida	385	49
Echiurida	4	2
Sipuncula	15	3
Pogonophora	13	5
Mollusca	106	43
Arthropoda	185	40
Bryozoa	1	1
Brachiopoda	2	1
Echinodermata	39	13
Hemichordata	4	1
Chordata	1	1
TOTAL	798	171

Properties Linked with Marine Species and Phyla

There are fewer known marine species, by about seven to one, than for the land, even though some intertidal and deep benthic areas and coral reefs possess equal species richness, as tropical forests. There are clearly many more orders and phyla (i.e. higher taxonomic levels representing broader genetic variation than species) in the sea. Moreover, important phyla are exclusively marine (endemic), for instance, Echinoderms, Ctenophors, Chaetognaths, Pogonophors, Brachiopods (Fig. 1), and most of those in benthic environments (many of them pass through a planktonic stage during their life history). If plants and protista are also considered, 88% of all phyla are found to be marine. For the moment, the functional role played by the vast taxonomic territory of classes, orders, families, and genera is virtually ignored. The utilisation, whenever possible, of cross breding experiments may reveal - as it has been shown in certain groups - an unexpected range of inter- and intraspecific differentiations affecting a multiplicity of characters such as degree of polymorphism, concealed genetic variation, sex determination and reproductive strategies. In certain copepods, for instance, the detection of reproductive barriers has led to the recognition of increasing numbers of sympatric or allopatric sibling species groups and closely related species within each genus (Battaglia and Beardmore 1978).

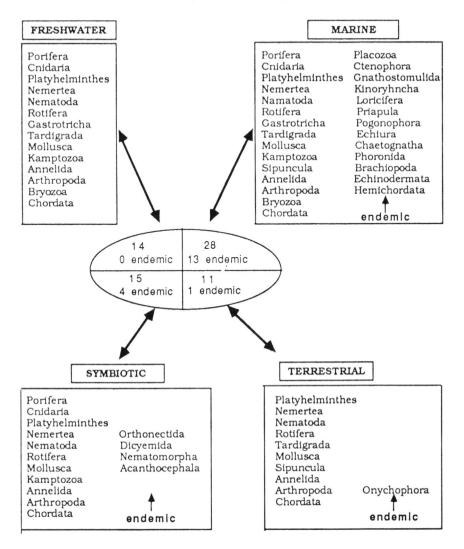

Figure 1. Distribution of Animal phyla by habitat. Numbers in central oval indicate total number of phyla that occur in each category (from Grassle *et al.* 1991).

Reservoirs of high marine diversity

Coral reefs are known as the most diverse communities in the sea, and they are most extensive in the western tropical regions of oceans, particularly the Caribbean and Indo-West Pacific. According to Jackson (1991), the question why coral reefs contain so many species remains one of the central questions of ecology despite more than a century of intensive fieldwork and theoretical investigations.

Little-explored marine habitats are a major source of novel discoveries of unknown high diversity reservoirs worldwide. For example, the number of marine fungi has risen from 209 species in 1979 to 321 in 1990 (Kohlmeyer andVolkmann-Kohlmeyer 1991). Marine microorganisms, i.e. algae, bacteria, fungi (including yeasts), protozoa, viroids and viruses are major contributors in biogeochemical cycles (Hawksworth and

Colwell 1992). Assessment of their biological diversity has been especially neglected. Fishes are the most abundant and diverse and poorly known group of vertebrate animals. The number of living fish species may approach 28,000 and this represents about half of all the known vertebrates species; of these 3/4 colonizes marine habitats. Soft-bottom macrobenthic communities, under temperate latitude, can be characterized by unusual high species diversity: e.g. in the Roscoff vicinity (Bay of Morlaix, Western English Channel), more than 1000 macrofauna species were identified from an area of 2.5 km2 (Gentil and Dauvin 1988).

Recent discoveries made in the deep sea, the deep hydrothermal vents, the sea ice biota and the almost unknown world of microorganisms have shown that present estimates of number of species and phyla may be underestimated by several orders of magnitude; they may represent reservoirs of biological diversity rivalling that of coral reefs and of tropical forests.

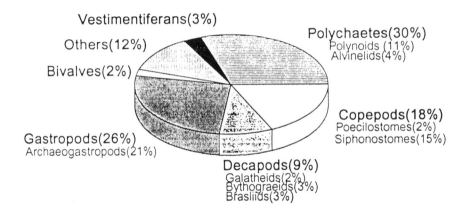

Figure 2. Major Taxonomic groups in the hydrothermal vent fauna (combined vent fauna sites of Mid-Atlantic Ridge, East Pacific Rise, Juan de Fuca, Explorer and Mariana). The total number of species is 193 (from Dover 1990).

Deep sea. Recent samples from the deep-sea floor in the North Atlantic suggest a previously unimagined richness of species (Grassle *et al.* 1990). From an area of about 21 m2 (the size of a small room), 798 species were identified representing 171 families and 14 phyla (Table 1). 460 of the species were new to science! Extrapolation suggests there may be millions of species, giving the deep sea an animal diversity of the same magnitude as the tropical rain forests. Echinoderms, sponges, pennatulids and gorgonians are among the most abundant epibenthic taxa of the deep sea floor. If global warming causes Arctic waters to heat up the effects on the deep benthos could eventually be devastatory. Simply because of the huge surface area, changes in deep sea communities could in turn have significant impact on global geochemical cycles.

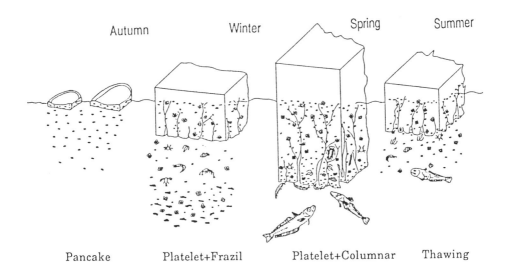

Figure 3. Seasonal colonization of sea ice by various kinds of organisms in the Southern Ocean (from Spindler 1990 and Hempel 1991).

Hydrothermal vents. Deep sea hydrothermal vents are unique environments characterized by their local insularity, global distribution. The taxonomic novelty of symbiotic and non-symbiontic invertebrates associated with the vent habitat is remarkable. More than 190 new species have been described. Records suggest a total 200 other new species. Species representation is strongly biased towards a few groups including polychaete worms, copepods, limpets (Fig 2). There is a high degree of endemism (Dover 1990).

Marine life of the sea ice biota. Seasonal ice of the Southern Ocean supports a distinctive biota consisting of representatives of many taxonomic groups of organisms, which spend all or part of their life cycles there (Horner 1985) and in part resembling to benthic systems. Most important are the ice algae (mainly pennate diatoms) and heterotrophic microorganisms (bacteria and lower fungi, ciliates, foraminifera, nematodes, turbellarians, polychaetes, amphipods and cyclopoids and harpacticoid copepods (Spindler 1990). The sea ice biota exists for only about eight months, from its formation in March to its disappearance in November (Fig. 3). How it is colonized by the various kinds organisms is not at all understood (Hempel 1991).

Picoplankton and microbial communities. The recent discovery of very small planktonic cells, called picoplankton has major implications for the primary production and turnover in oceans. In the size range of 0.2 to 2 microns, with live weights on the order of a picogram, the picoplankton comprises pico-eucaryotes, cyanobacteria and prochlorophytes, and it is recognised as the largest numerical component of the phytoplankton in the ocean and may play an important role as a sink for atmospheric carbon dioxide. The discovery of picoplankton had to wait the flow cytometry technology (Fig. 4). Flow cytometry is new to oceanography and register and amplify fluorescence of living cells made to flow past a region of excitation by monochromatic light emitted by a laser beam. Several thousands cells per minute can be analysed for size pigment and

DNA contents. Immunofluorescence techniques (specific antibody against whole cells) permit to distinguish sero groups probably corresponding to different species (Campbell *et al.* 1983, 1989).

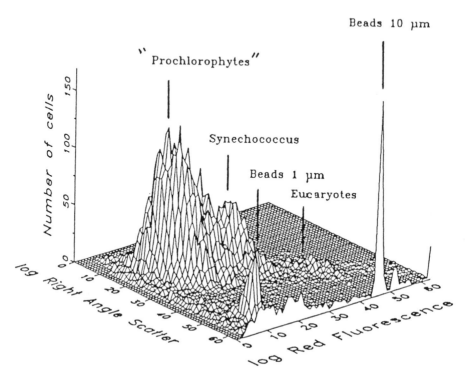

Figure 4. Cytogram obtained with a ship-board flow cytometer in the surface waters of the northwestern Mediterranean Sea. It shows the three main populations of picoplankton (size 0.2-2 um): pico-eucaryotes, cyanobacteria (*Synechococcus*) and prochlorophytes (from Vaulot *et al.* 1990).

The use of molecular probes to characterize marine biodiversity

To what extent molecular biology and analytical biochemistry, and molecular genetics, could not be of any use in monitoring marine biodiversity? Recent methodological developments, such as advances in the application of recombinant DNA (rDNA) techniques, in conjunction with comparative physiology, will expand our capabilities to reveal which character may be adaptation to the environmental conditions.

Current investigations of phylogenetic systematics using molecular cladistics begin to incorporate the vast taxonomic territories of classes, families, genera which where virtually ignored. New molecular techniques are being developed in phylogeny reconstruction and marine biotechnologies to forge new links between biodiversity and ecology.

A variety of recent investigations have demonstrated that the algae are not a natural group. All of these analyses agree that many groups of photosynthetic organisms included in the algae are more closely related to protists or fungi than to higher plans or to each other (Gibbs 1990). Cladogram on figure 5 is based on ribosomal RNA sequence comparisons, it shows that algae are polyphyletic. Euglenoids evolved very early and are related to Trypanosomes. The dinoflagellates evolved late and are related to ciliates. The brown algae and Chrysophytes are related to each other and to Oomycetes. Higher plants and green algae form a natural group. It is remarkable that organisms with chloroplasts appear on different branches of the tree, indicating that different groups have acquired chloroplasts independently. Other investigators have shown that the arrangement of genes coding for the large and small subunits of ribulose-1,5-bisphosphate carboxylase/oxygenase (RUBISCO) differs among lineages of photosynthetic organisms. The only link between Dinoflagellates, Cryptophytes and what could be called Chromophytes stricto sensu (Rhaphidophycae, Chrysophyceae, Haptophyceae, diatoms and brown algae) is that their plastids have probably evolved from red alga-like ancestors (Assali *et al*. 1991).

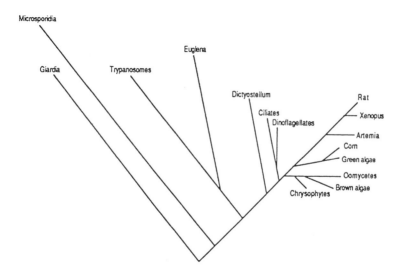

Figure 5. Eukaryotic phylogenetic tree based on subunit ribosomal RNA sequence comparisons (from Gibbs 1990).

Bacteria, yeast, and viruses die off in sea-water need to be revisited, in light of new work showing that bacteria undergo starvation and entrance into dormant stage, in which they are viable, but nonculturable. A recent area of endeavour has been the use of rRNA sequences to characterize marine microbial diversity in natural ecosystems that cannot be grown in culture (Giovannoni *et al*. 1990). These phylogenetic characterizations are more than an exercice in taxonomy. Giovannoni *et al*. (1990) have used this approach to study the species diversity and populations structure of picoplankton in ocean waters. They identified genetic markers that can be used to examine the distribution and genetic diversity in Sargasso Sea bacterioplankton, with nucleic hybridization probes in routine studies. The microbial populations are adapted in the Sargasso Sea, a central oceanic gyre, under extremely low nutrient conditions. The analysis fully support the widespread view that microbial ecosystems contain novel, uncultivated species: a SAR11 cluster appears to be a significant component and a second cluster of lineages related to the oxygenic phototrophs - i.e. cyanobacteria, prochlorophytes and chloroplasts - was also observed, with unexpected genetic variability within the clusters. The genetic diversity indicated by this important study suggests that rRNAs may provide a powerful tool for analysing prokaryotic population biology, as well as for identifying community numbers, such as different clusters of lineages. For planktonic bacterial populations, viral infections may be a source of selective pressures contributing to the divergence of lineages (Proctor and Fuhrman 1990). A close association of Vibrio cholerae with planktonic copepod populations in eastuaries has recently been described (Tamplin *et al.* 1990).

The ecological message is clear. The new culture-independent method is beginning to reveal some of the unexplored diversity within the marine microbial world.

The Role of Ecological Processes in Shaping Patterns of Marine Biodiversity, in Space and Time

This topic addresses questions on the importance of biological diversity in controlling ecological processes, to discover to what extent patterns of biological diversity are important in determining the behaviour of ecological systems.

Evolution and genetic variation within marine populations

Physiologists generally tend to assume that the traits that they examine represent adaptations to the environment. Similarly, most ecologists have tended to assume that the organisms found in a particular location are present because they have the highest fitness. This is essentially a neo-Darwinian view of the world that has been vigorously attacked over the last decade. Cladistic analyses and some statistical techniques, such as nested analyses of variance, can help to reveal which characters may be adaptation to local conditions and those that are simply artifacts of an organism's evolutionary history. These observations have predictive value for plant and animal ecologists (Wanntorp *et al.* 1990).

We might expect two populations living in environments of differing salinity or temperature to differ genetically due to adaptation to local conditions. Mussels (*Mytilus edulis*) collected over a wide range of ambient salinities show inhibition of oxygen consumption when salinity is changed from that of ambient conditions of their native habitat. The genetic difference involved here is unknown but potentially great. Although physiological differences among populations may have a genetic basis, individuals may irreversibly acclimatize to their native environment. Therefore a genetic difference cannot be inferred simply because individuals collected from two different environments show different physiological response. Active research is now developing on protein polymorphism, genetic polymorphism, physiological adaptation, with a focus on the responses of marine species to environmental stress (tolerance, resistance and adaptation). The incorporation of new molecular techniques is necessary to better characterize genetic differences.

Mytilus species are distributed worldwide in the littoral zone of temperate and cold oceans. Many species have been described, but there is little consensus as to which are actually distinct lineages. Allozyme studies suggest that at least three species (*M. edulis*, *M. galloprovincialis*, and *M. trossulus*) inhabit the Northern Hemisphere (McDonald and Kochn 1988). Both *M. edulis* and *M. trossulus* are found in the western North Atlantic near the Gulf of St. Lawrence and may hybridize in areas of overlapping range (Vario *et al*. 1988). Analyses of mitochondrial DNA from 16 populations of the *Mytilus edulis* group showed, within each population, an unusually high frequency of heteroplasmy and divergence from mtDNA sequences. Of the 150 *Mytilus* individuals, 85 were heteroplasmic (Table 2, Hoeh *et al*. 1991). It is hypothesized that biparental inheritance of mitochondrial can occur in *Mytilus*. Because only highly differentiated mtDNA types were observed in heteroplasmic *Mytilus* from the Gulf of St Lawrence and because hybridization between *M. edulis* and *M. trossulus* is possible, biparental inheritance of mtDNA may be a hybrid phenomenon (Hoeh *et al*. 1991).

Table 2. Geographic distribution of heteroplasmy in *Mytilus* (from Hoeh *et al*. 1991).

Region	Populations sampled	Number of individuals	Number (%) heteroplasmic
Western North Atlantic	4	57	31 (54%)
Eastern North Atlantic	5	52	34 (65 %)
Western South Atlantic	3	21	9 (43 %)
Western North Pacific	1	6	3 (50 %)
Eastern North Pacific	1	2	1 (50 %)
Western South Pacific	2	12	7 (58 %)
TOTALS	**16**	**150**	**85 (57 %)**

Biogeographic patterns in marine biodiversity

The marine environment is the only uninterrupted, internally moving, three-dimensional environment. In principle, each organism can reach all places of the oceanic world. The biogeographic patterns of marine flora and fauna at the general oceanic scale are interesting to consider as they are the product of the geological history of the oceans, the continental barriers, the current patterns and limitations of survival of individual species and populations by both biotic and abiotic conditions.

Oceanic currents and plankton distribution. The general distribution patterns of planktonic species has been studied thoroughly (Spoel and Pierrot-Bults 1979; Spoel and Heyman 1981). Plankton, per definition, depends in its movements on water circulation, currents can transport easily specimens from their reproductive original area. The total distribution of a planktonic species consists of a reproductive range, a non-sterile expatriation range and a sterile expatriation range. Moreover, many species show vertical migration which gives their ranges a vertical component.

Biodiversity and ecology of marine parasites.   The term parasite encompasses a variety of organisms including viruses, bacteria, invertebrates, vertebrates, and plants. The parasites of many seas have not been sufficiently examined and large geographic areas are virtually unknown (Rohde 1993). Knowledge of their biology is crucial to any study on the role played by marine parasites in controlling the diversity of free-living species, and *vice-versa*. Of significance too is the role played by parasites in the regulation of host populations, and in ecosystem function.

Among the Meterozoan parasites of pelagic hosts, the most important are the platyhelmintes (trematodes, cestodes) of which the hosts are fishes. Their larval stages cercariae, metacercariae, coracidia, procercoids, and plerocercoids, develop in planktonic invertebrates such as Ctenophora, Polychaeta, Chaetoganath and Custacean, or they are free living in the plankton. The parasitic dinoflagellates (peridinians) are particularly numerous, occuring in broad ranges of organisms from Protista to fishes and showing morphological or biological adaptation in relation to parasitism from epibiontism to endoparasitism (Cachon and Cachon 1987). There are relationships between latitudinal gradients in species diversity of hosts and parasites. Data from tropical seas is incomplete. Considering Monogenea, the data base is still small but it seems that the Indo-Pacific has generally more species and more endemics than the Atlantic Ocean (Rohde 1993). Many species of marine animals migrate over long distances, and changes in the composition of their parasite faunas may occur during such migrations. The value of branchial Monogeneans for species identification of their hosts has been demonstrated (Euzet *et al.* 1989). Changes in the parasite fauna may be small or non-existent if host populations migrate between different seas, and this facilitates use of parasites as biological markers. One example is to use parasites for tracing nursery grounds and migration of fishes (Groot *et al.* 1989).

Diversity gradients. Three principal gradients for species distribution are important in the oceans: (1) depth gradients, (2) salinity gradients and (3) latitudinal gradients.
(1) Depth gradients. In aquatic environments, both marine and fresh water, diversity generally decreases with increasing depth. Because this is also usually a gradient of decreasing temperature, it invites comparisons with elevational gradients on land. However, the comparisons are misleading because light and seasonality also decrease and pressure increases with water depth. In most bodies of water, depth patterns are further complicated by the fact that the greatest diversity occurs not in the sallowest waters, but some distance below the surface (Fig. 6, Rex 1981).

(2) Salinity gradients. In several coastal regions marine organisms encounter local gradients of decreasing or increasing salinity. Almost invariably, diversity declines as the concentration of dissolved solutes deviates from normal sea-water. Note in figure 7 that the pattern of estimated diversity is bimodal, with distinct peaks corresponding to fresh water (inf. 2 g/l) and sea-water (30-40 g/l). One consequence of these relationships is that estuaries and other brackish habitats where sea and fresh waters meet typically have low species diversity, although these waters may be highly productive and support dense populations of some species (Kinne 1971). Estuarine systems, including true estuaries, coastal lagoons, deltas, mangroves confront potential colonizing organisms with severe osmotic problems and so some to support highly adapted communities (Lasserre 1976b).

(3) Nutrient gradients. The ocean is a solution of nutrients in the dissolved state, with a depleted surface layer due to the uptake by living forms in the euphotic zone. Exchange processes, such as upwelling and wind mixing, may balance the loss of surface nutrients to greater depths via sinking carcasses and zooplankton fecal pellets. The nutrient regime

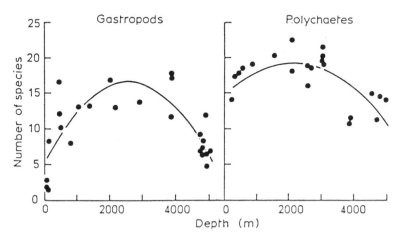

Figure 6. Variation in species richness with ocean depth in two groups of benthic marine invertebrates, molluscs and polychaetes (from Rex 1981 and Levinton 1982).

is generally regarded as one of the major factors defining productivity in the oceans (Ryther and Guillard 1959). The spectrum of limiting nutrients has been implicated as a factor controlling species diversity of the phytoplankton community (Tilman 1977).

In general if follows that diversity generated in the phytoplankton community (which ranges in species number and size distribution over many orders of magnitude) can also decide species diversification in the zooplankton community, and eventually also at higher trophic levels. Tilman has considered the co-existence of phytoplankton species as being the result of different requirements of the various algal species for the whole spectrum of available nutrients in the sea. For coral reefs, a similar observation to Tilman's has been made by Birkeland (1988). Birkeland has considered the nutrient field to be the analogue of moisture in the terrestrial environment. His general hypothesis is that the plants and animals follow a gradient of decreasing nutrient input (i.e. from land or upwelled areas). Species richness from heterotrophic suspension feeders to benthic plants and finally phototrophic animal symbionts increased in a direction away from the nutrient supply.

It appears that there is no general consensus of what effects are caused by changes in the nutrient gradient - in fact this may be something that changes within the habitat. Effects documented in the Indo-Pacific may not apply to the Carribean. Thus one useful avenue of research might be to test the effect of changes in experimental conditions in high diversity environments. While the question is scientifically interesting, it is also of practical importance in tropical ecosystem management.

The difficulty of many observations in the marine habitat is that water moves around and no single observation can be repeated on exactly same water body. One solution to this problem is to use controlled ecosystem which are large enough to contain some part of the diversity of the relevant ecosystem which is small enough to contain

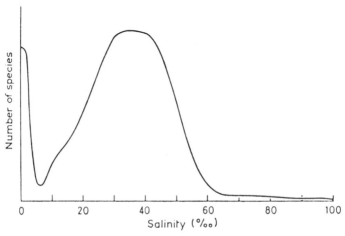

Figure 7. Number of species in relation to salinity. Note bimodal pattern with distinct peaks corresponding to fresh water (inf. 2 $^o/oo$) and seawater (3-4 $^o/oo$) (from Remane and Schlieper 1971 and Kinne 1971).

consecutive stages in the growth stages of the different species within the confined environment. Controlled experimental ecosystems have been developed largely for pelagic and benthic systems and for temperate near-shore communities (Lalli 1990). Their adaptation to tropical environments seems a natural evolution of this experimental technique.

(4) Latitudinal gradients. One of the most well-known gradient is an increase of species richness from high to low latitudes in continental shelf. This pattern has been recorded for planktonic foraminifera and mollusks (Fig. 8, Stehli *et al.* 1967). Few detailed studies have compared species richness to habitat diversity at different latitudes. Spight (1977) examined prosobranch gastropods diversity in beaches of Washington and Costa Rica, he found that the regional lists contained five times more species, in Costa Rica, than in Washington. The tropical community seems more diverse because there are many available substrates with no temperature counterparts, not because competing species have divided up the available resources more finely than in the temperate zone (see also Levinton 1982). The large number of small species of caridean, brachyuran and anomuran, at low latitudes, offers an explanation of the higher species diversity found in the tropics. Changes in the size composition of decapod crustaceans of the continental shelf with latitude are shown to be due mainly to changes in the abundance of small species (Steele 1988).

More generalization of latitudinal species-diversity gradients have been proposed, but the general evidence is weak and contradictory, even if the idea is now entrenched in many textbooks of marine ecology. The papers of Sanders (1968), among others, predicted highest species richness per habitat in the tropics. They hypothesized that physically stable marine environments, such as coral reefs and soft bottom sediments, favour the development of more diverse, species-rich communities in comparison with young or stressed regions where physical factors dominate in regulating community diversity.

More strictly comparative studies using reliable sampling strategies and appropriate observation scales, sampling, and modelling (Frontier 1985) are needed. Warwick and Ruswahuri (1987) could find no difference in macrobenthic diversity between sites in Java and in the U.K. For seaweeds (macro-algae), the most diversified systems are in the temperate, nearshore areas of California, Japan, southern Australia, the northern Atlantic, and the Brittany coast of France. Contrary to general opinion, coral reefs are often less diverse than are neighbouring level-bottom environments, and the low-diversity communities may be at least as stable as the high-diversity communities (Jackson 1972).

Many more comparisons based on more strict comparisons will need to be made before any generality can be achieved. The contention that tropical intertidal communities are, on the average, subjected to greater environmental stress than temperate organisms should be also considered. (Alongi 1987).

Is biogeography relevant to studies of scale changes in marine systems? The high diversity coastal ecosystems contain many pelagic or benthic (or both) species. These species range widely over several order of magnitude in both size and abundance. Thus, the study of the total pelagic and/or benthic communities is logistically a very complex problem. There is no way to sample whales and picoplankton within the same time/space domain. Larger species tend to be more mobile, their home ranges are more extensive, and individuals can exert considerable control over their location. Very small organisms are unable to control their distribution in space by their behaviour, but in many species their reproductive rates are rapid enough for them to exploit quite fine-scale events. Hence, the problem of discriminating between population and community changes which are responses to coarse scale events, such as those generated by interannual and decadal variations in climate, and those which are caused by shorter term variations induced by "ocean weather", is far from simple.

Biogeographical provinces delineated in the oceanic pelagic realm tend to correspond closely to the patterns of the general circulation of ocean currents, which are in part driven by the winds and so are likely to be responsive to shifts in global climate. The boundaries to biogeographical provinces are often ill-defined, because individuals tend to spread beyond the distributional limits for reproductive success, where they may persist without breeding. Recently, Angel (1991) re-considered the relationship between time and space variability in oceanic zooplankton biomass; he stressed that the temporal scales of interest are those 1 year, which are associated with spatial scales of thermohaline and climatic transients and the biogeographical ranges of species and communities. Few individual species have distributional ranges which correspond exactly to the chemico-physical boundaries observed within the environment. A species' potential range is determined by the physiological limits to its survival, i.e. extreme of temperature, pressure, light or chemical concentrations. Most species, however, fail to expand into their full potential range, because they are limited by biological interactions (such as competition or mutualism).

Active selection of benthic substrates at the time of larval settlement has only recently been shown to occur under realistic flow conditions (Butman *et al*. 1988). The effects of hydrodynamic processes and chemical cues on larval settlement were tested simultaneously under controlled conditions (Pawlik *et al*. 1991). The experiments were conducted with larvae of a reef-building tube worm. The results suggest that although physical processes are largely responsible for the delivery of larvae to potential habitats, larval behaviour in response to flow conditions may facilitate substrate contact and, more generally, play an important role in larval recruitment to the benthos.

Failure to consider the effects of scale has been a major source of confusion in theories to explain coral reef diversity. Patterns of coral distribution are more predictable at larger scales (Jackson 1991), however, there has been little focussed study of the causes of coral distributions and diversity, except at the smallest spatial scales (e.g. one-square-metre quadrats). According to Jackson, there is a pressing need for more large-scale, descriptive, and experimental studies of coral distribution and ecological processes within single habitats and across environmental gradients (see also Birkeland 1988).

Role of marine biodiversity in the global carbon cyle

Biological pumps and the global carbon cycle. The oceanic world is the largest sink of the planet: There is 15 times more carbon in the oceans than occurs on land (in plants, animals and sediments) and 50 times more than in the atmosphere. Moreover, the oceans provide an enormous potential sink for man-made emissions of the so-called "green house". Exchanges between the two atmospheric and oceanic reservoirs are very intense: all atmospheric CO_2 transits through the ocean in 8 to 10 years, the part of transit through the marine biosphere reaches 30 to 50%. The marine biosphere is characterized by low instantaneous biomass (0.5% of the terrestrial biomass), however, mechanisms producing CO_2 fluxes are very rapid in the ocean (residence time for C in the marine biosphere is inferior to one month, against 25 years for terrestrial biosphere): the fluxes transiting by terrestrial biosphere and those transiting by marine biosphere are of the same order of magnitude. In a non disturbed "natural" environment (no human perturbation, that is in a near steady state), the exchanges with atmosphere are close to the global balance: the CO_2 solubility increases when temperature decreases, that is CO_2 is emitted in the atmosphere at the tropical levels (low latitudes), and, on the contrary, CO_2 is dissolved at high latitudes. The "biological pumps" act everywhere at any time to modify the efficiency of the "physical (or thermodynamic) pump". The human activity has modified this equilibrium: oceanographers recognises that ocean may have stored 30 to 50% of the anthropogenic CO_2 production to the atmosphere.

In the oceans, biological CO_2 pumps are vertically-directed, promoting the transfer of carbon from surface to deeper water. Several pathways are involved: (i) the sinking of plant and animal debris, containing both organic carbon and calcium carbonate; (ii) active downward transport, brought about by the feeding and excretory behaviour of migratory zooplankton ("gut flux"); and (iii) the downward advection and diffusion of dissolved organic carbon, produced by decomposition processes in the upper ocean. The net effect of such biological activity is to reduce the partial pressure of CO_2 in surface waters, causing the drawdown of CO_2 from the atmosphere.

The rate of return is subsequently determined by the depth distribution of respiration and decomposition processes, and by physical factors influencing carbon transport in the water column. Quantifying the importance of these carbon circulation processes in different oceans, and for the world as a whole, is a major aim of the Joint Global Ocean Flux Study (JGOFS), a core project of the International Geosphere-Biosphere Program (IGBP). A new IGBP core project, i.e. the Land Ocean Interaction in the Coastal Zone (LOICZ) will focus on similar aspects.

Photosynthesis is limited to the sunlight, upper ocean whilst respiration occurs throughout the water column. The organisms involved are mostly near-invisible: several millions algae, bacteria, protozoa and other forms of life inhabit a breaker of water scooped from any ocean, unseen except by microscope. Yet together such species have changed the Earth: over geological periods of time, marine plankton and benthos have been responsible for a vast accumulation of carbon in the oceans and in the sediments, altering atmospheric composition and the heat budget of the planet. In these phenomenons, the qualitative role played by marine biodiversity - which species, which community, which ecosystem play important roles? - is almost unknown. The following example is, however, worth of interest.

Adelie penguins and global change in the Antarctic ocean. A recent study underlined that Adelie penguins and marine mammals are a vital link in the Southern Ocean food web. However, their respiration of CO_2 may represent a significant inefficiency in the storage of fixed carbon in the ocean. This phenomenon may affect current models of the global ocean-atmosphere carbon flux (Huntley *et al.* 1991).

Primary production in the Southern ocean is approx. 3.5 gigatons of carbon per year, which account for nearly 15% of the global total. The presence of high concentrations of nitrate in Antarctic waters suggests that it might be possible to increase primary production significantly and therefore alleviate the net accumulation of atmospheric carbon dioxide. The result would be to increase the activity of the biological carbon pump and to sequester atmospheric CO_2. However, little consideration has been given to the ultimate fate of this carbon as it is transferred through the Antarctic marine food web.

Huntley *et al.* (1991) have modeled the annual mean flux of carbon through this food web, considering penguin and seal mammals respiration, a term previously neglected. Approximately 2/3 of Southern Ocean bird biomass is composed of the penguins, while they eat krill exclusively, other bird species also rely largely on krill.

Moreover, almost 90% of the Antarctic sea biomass is composed of the crabeater seal whose diet is 90% krill; apart from the fur sea whose diet is 30% krill, and the leopard sea, which preys upon krill, birds, and seals; other species of Antarctic seals depend primarily on fish and squid. The structure of the model reflects available knowledge of trophodynamic relations in the food web of the Southern Ocean (Fig. 9).

The CO_2 produced by respiration of penguins and seals is presumed to enter the atmosphere directly, without opportunity for gas exchange with sea-water; CO_2 respired by all other species is assumed to go directly into sea-water. On the basis of this model, as much as 0.86 gigaton of CO_2 per year could be returned to the atmosphere through Antarctic birds and mammals respiration. It is in such near shore regions of the Antarctic that the greatest atmosphere-ocean disequilibrium in pCO_2 has been observed for 1 to 2 months during the spring bloom of phytoplankton. The authors suggest that similar determinations be made for birds and mammals in other oceans to assess their global role in the biological carbon pump (Huntley *et al.* 1991).

The Importance of Marine Biodiversity in Controlling Ecological Processes

Marine and terrestrial ecosystems differ significantly in their functional responses to environmental change. In terrestrial systems, environmental variability is large at both short- and long-term periods and physical boundaries between ecosystems are well marked. This suggests that terrestrial organisms and ecosystems have developed internal mechanisms to cope with short-term variability and, in turn, may serve to minimize effects of long-term variations (Steele 1985). In contrast, in marine systems, physical variability is small and extends over very long time scales due to the large thermal capacity of the ocean and the long period exchange between deep and near-surface waters (100 to 1000 years). Physical boundaries between system are weak. Therefore, marine ecosystems are more vulnerable to large scale environmental changes because they do not have the internal adaptability inherent in terrestrial systems. At time scales of several years to decades, marine systems are more closely coupled to their physical environment than are terrestrial systems (Steele 1985). As a consequence, most marine systems may respond to rapid changes in their overall environment and if they may be much more sensitive to alterations in their environments, they may also be much more adaptable (Steele 1991).

Responses to environmental change. The effects of physical processes on particular populations, during the early life stages, can be explained, if one combine the classical terrestrial mechanism: competition for space, and the classical marine mechanism: larvae carried by currents. In the ocean, larval dispersal provides recruitment to areas where adults are not present. This broad larval dispersion may allow populations in one area to live dangerously if there is always the possibility of recruiting larvae from elsewhere (Steele 1991). On the other hand, observed changes at the community level have not been explained. One well described example is the changes in the North Sea food web (Steele 1974). During the last three decades, there have been changes in the abundance of herring and some of the planktonic species it consumes. These changes may be related to climate changes (Aebischer *et al.* 1990). However, during the same period, there have been marked increase in several demersal fish species such as haddock. These fish occupy a quite different position in the foodweb than herring (Fig. 5). It has not proved possible to explain these switches in terms of particular causal processes in the food web.

There are many examples of striking switches in community structure. Large switches can last several decades and have major economic consequences. The Russel cycle in the English Channel (Southward 1980) consists of marked changes in the herring and pilchard populations during the last 80 years. Similar switches of seaweeds and macrobenthos have been made in the Roscoff area, (Western English Channel), over several decades, apparently in conjunction with temperature fluctuation of approximately 1/2 °C (unpublished data).

In situ experiments in conjunction with continuous measurements of environmental variation, on interannual or decadal time scales are needed to understand how the pattern of variability translates into marine functional diversity (Steele 1991). In coastal water, the interpretative problems are exacerbated by anthropogenic influences such as changes in nutrient status through waste discharge and agricultural runoff, and impact of fishing which both upsets the competitive balance by removing top predators and can cause physical disruption of benthic habitats.

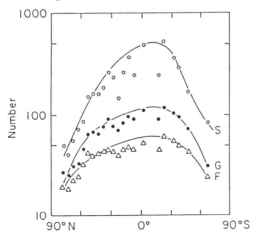

Figure 8. Bivalve mollusk species richness as a function of latitude. S = species; G. = genera; F = families (from Stehli *et al.* 1967).

The 1982-83 ENSO sea warming event. The phenomenon called El Niño Southern Oscillation (ENSO) emphasises the fact that atmospheric variations can mediate changes in the oceanic environmental on a global scale, with major biological consequences, throughout the tropical oceans. During the extraordinary strong 1982-82 ENSO event, all zooxanthellate reef coral species in the equatorial eastern Pacific experienced some degree of bleaching and mortality during the prolonged (5-6 months) sea water warming (2°C to 3°C) above normal period. Recently, Glynn and Weerd (1991) reported that two of the 12 Panamian coral species (Millepora) were eliminated suddenly from their former ranges: The two hydrocorals species have not been seen alive in Panama since 1983. This may represent the first documentation of an extinction (total loss of genetic

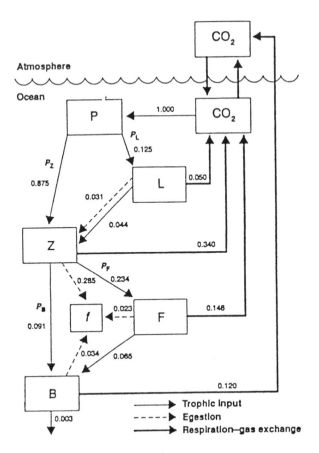

Figure 9. Antarctic Marine food web structure and model output values for an initial photosynthetic value of 1 unit,. P = primary production; Z = macro-zooplankton (mostly krill); L = microbial loop (ciliate, bacteria); F = fisches and squids; B = birds (penguins) and mammals (seals); f = fecal matter. Respiration of carbon dioxide by Adelie penguins and seals (B) may transfer as much as 20 to 25 percent of photosynthetically fixed carbon (from Huntley *et al.* 1991).

diversity) and an extreme range of reduction of coral reef species, perhaps a harbinger of future extinctions should habitat destruction and coral reef bleaching events continue (Glynn and Weerd 1991). If present species eliminations associated with ENSO warming can be extended to possible future global warming, reef-building corals with populations confined to small geographic areas could succumb to slight but sustained sea warming.

Along the peruvian coast, the 1982-83 ENSO event has marked effects, positive and negative. Soft bottom macrobenthos at 34 m depth in Ancon Bay, Peru, responded positively (Tarazona *et al.* 1988, Fig. 10). In the first months of ENSO (October-December 1982) the number of species increased from 6 to 24 due mainly to recruitment of polychaete worms. Population densities and biomass increased markedly with the onset of ENSO and oscillated greatly for the next twenty months. Species diversity nearly tripled by the end of February 1983. On the contrary, many species of crabs, molluscs and sea-urchin suffered mass mortalities while others recorded post-event increases (scallops, octopus, purple snail) or extended their distribution (barnacles, shrimps, lobsters). The result was generally beneficial for the fisherman, particularly those exploiting the enormous populations increases of the local scallop (*Argopecten purpuratus*).

Responses to stress and to pollution. Benthic assemblages in the low latitudes may be subjected to greater stress than their temperate counterparts (Moore 1972). Physiological studies indicate that tropical organisms are generally as intolerant to temperature and salinity stress as are temperate and boreal organisms. Mayer (1914) found that tropical marine organisms can withstand cooling better than heating, indicating that lower temperatures constitute less stress than higher temperatures.

Survival and growth of three species of the marine ciliate Euplotes (temperate, tropical and Antarctic sea-ice species) were studied at different temperatures in laboratory cultures (Lee and Fenchel 1972).The Antarctic species survived and grew between -2°C and 10°C (*Euplotes antarcticus*),the temperature (*E. vannus*) and tropical species (*E. balteatus*) survived between -2° to about 30°C and between 5 and 40°C respectively (Fig. 11). Ecophysiological studies on marine meiofauna living in the temperate sandy beaches (Lasserre 1976a), showed adaptive plateaus for respiration in relation with the intertidal distribution of the different species, showing distinct temperature tolerances. In tropical coral beaches (Wieser and Schiemer, 1977) found that the most abundant nematode species exhibited distinct temperature tolerances in relation with their distribution in the upper beach on in the lower beach levels.

Investigating the impact of environmental stress at various levels of biological organization, from ecosystem, through population to individual, physiological and genetic levels, involves consideration of its use as a probe of ecological systems and evolutionary processes. The requirement for careful evaluation of the selective forces represented by particular environmental stress is a challenging problem when attention is directed to comparisons between species (Bayne 1985). For Koehn and Bayne (1989), stress is considered as an environment change that results in reduction of net energy balance (i.e. growth and reproduction); they emphasize on Mytilus individual differences in net energy balance and the interrelationships among genetic heterozygosity, and they consider how genetic/environemental interactions can define individual responses to environmental extremes.

The range of marine species, however, on which we have a detailed knowledge on effects of stressors is very limited. The extrapolation of effects at the individual level to effects at the population level is made difficult by large temporal fluctuations in abundance of many organisms. Monitoring can measure this variability but is insufficient to predict the potential impact of most stresses. Moreover, unless stressors can be shown as having effects at the population level there are unlikely to be ecologically significant (McIntyre and Pearce 1981). Natural mortality or larval phases is usually high and detection of effect of stressors at the individual and population level against a background of high natural mortality is difficult. A very good review of how stressors affect different marine benthic assemblages has been made by Gray (1989). Many examples of reduction in diversity in polluted areas are known. Species diversity is more strongly influenced than density (Moore *et al*. 1987), however, there is little doubt that significant reductions in diversity are found rather late in the sequence of responses to stressors.

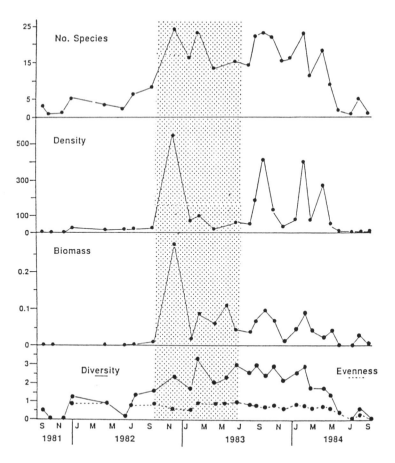

Figure 10. Number of species, density, biomass and eveness of macrobenthos at 34 m depth in Ancon Bay, Peru (September 1981 to September 1984). Shaded area corresponds to the El Niño period (from Tarazona *et al.* 1988).

One crucial question is whether or not the changes observed are causes by man-made stressors or are natural processes. Pearson and Rosenberg (1978) have given a general model of effects of organic enrichment of marine benthos (Fig. 11). The model predicts that along some organic enrichment gradient, the first changes are a slight increase in a number of species and a rapid increase in biomass, followed by a fall in species number and biomass. The dominance by the opportunists is almost the last stage in the retrogression of the system. Data showing such tendency have been obtained experimentally for meiobenthos subject to the Amoco Cadiz oil spills (Boucher *et al.* 1982).

One of the characteristics of marine benthic assemblages that are exposed to high stressor loads is the presence of species complexes. The studies on *Capitella capitata* (Grassle and Grassle 1976) show that from the Massachusetts coast there are not one but six genetically distinct species. Such complex of species dominates in organically enriched habitats. To explain why should there be species complexes in the most disturbed marine habitats, Grassle and Grassle (1977) suggest that in such groups the taxonomic unit that survives through evolutionary time may be a metaspecies where species are continuously being formed and becoming extinct.

Organic enrichment gradient

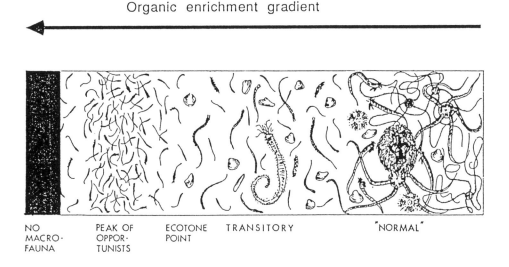

| NO MACRO- FAUNA | PEAK OF OPPOR- TUNISTS | ECOTONE POINT | TRANSITORY | "NORMAL" |

Figure 11. Diagrammatic representation of changes in abundance and species types occurring along a generalized organic enrichment gradient (from Pearson and Rosenberg 1978).

Conclusion

The investigation of marine biodiversity poses a considerable scientific and conservation challenge because of the great size and relative inaccessibility of marine ecosystems. Only now is marine biodiversity beginning to be understood, the implications of its different facets being appreciated. New reservoirs of high diversity have been described, from deep sea to polar oceans. Molecular biology techniques offer new links between biodiversity and ecology. Phylogenetic analyses and ribosomal RNA-based methods have led to discover large number of marine microorganisms, which have never been cultured. These methods reveal some of the unexplored diversity within the marine microbial communities.

We cannot expect too many similarities with terrestrial situation because the spectrum of environmental variation in the sea has a longer wave length over both ecological and evolutionary time scales than on land. Physical boundaries between ecosystems are less pronounced than for terrestrial systems, and marine ecosystems and organisms may have developed less robust internal processes to respond to the low magnitude environmental short-term variations. This would result in a reduced ability to respond to large scale environmental changes. Therefore, it may be apparent that changes in biodiversity are more rapid in marine ecosystems than in terrestrial ecosystems. However, some marine ecosystems may appear much more adaptable than individual species.

There has been little focussed study of the causes of latitudinal species distribution (including coral reefs) except at the smallest spatial scales (Ogden and Gladfeter 1983). There is a pressing need for more large-scale, descriptive and experimental studies of communities distributions and ecological processes within single habitats and across environmental gradients. The existence of biodiversity gradients (e.g. low versus high latitudes) is not clear and their functional significance should be reappraised. Also the relationships of species richness to such a process as productivity remain dubious. It is clear that a global plan is needed to better explain how human activity affects marine biodiversity and to evaluate the consequences of diversity changes, most notably in the coastal zone (Grassle *et al.* 1991). The scale of marine systems and the mixing, dispersion and the transport that occur in the oceanic medium require different thinking for investigative processes, as well as for conservation and sustainable management.

References

Aebischer N.J., J.C. Coulson and J.M. Colebrook, 1990. Parallel long trends across four marine trophic levels and weather. *Nature* 347: 753-755.

Alongi D.M., 1990. The ecology of tropical soft-bottom benthic ecosystems. *Oceanogr. Mar. Biol. Annu. Rev.* 28: 381-496.

Angel M.V., 1991. Variation in time and space: is biogeography relevant to studies of long-time scale change? *J. mar. biol. Ass.* UK 71: 191-206.

Assali N., W.F. Martin, C.C. Sommerville and S. Loiseaux-de Goer, 1991. Evolution of the rubisco operon from prokaryotes to algae: structure and analysis of the rbcS gene of the brown algae *Pylaiella littoralis*. *Plant Molec. Biol.* 17: 853-863.

Battaglia B. and J. Beardmore, 1978 (eds.). *Marine organisms: genetics, ecology, and evolution.* NATO Conf. Series. New York: Plenum Press.

Bayne B.L., 1985. Responses to environmental stress : tolerance, resistance and adaptation. In J.S. Gray and M.E. Christiansen (eds.), *Marine biology of polar regions and effects of stress on marine organisms*, pp. 331-349.

Birkeland C., 1988. Geographic comparisons of coral-reef community processes. In J.H. Choat *et al.* (eds.), *Proceedings of the 6th international coral reef symposium. vol. 1. Symposium executive committee*, pp. 211-220. Townsville, Australia.

Boucher G., S. Chamroux, L. Le Borgne, and G. Mevel, 1982. Etude expèrimentale dune pollution par hydrocarbure dans un microècosystème sèdimentaire. I. Effets de la contamination du sèdiment sur la meiofaune. In E.R. Gundlach (ed). *Ecological study of the Amoco Cadiz oil spill*, pp. 229-243. Report of the NOAA-CNEXO Joint Scientific Commission.

Butman C.A., J.P. Grassle and C.M. Webb, 1988. Substrate choices made by marine larvae settling in still water and in a flume flow. *Nature* 333: 771-773.

Cachon J. and M. Cachon, 1987. Parasitic Dinoflagellates. In F. J. R. Taylor (ed.), *The biology of dinoflagellates*, pp. 571-610. Oxford: Blachwells.

Campbell L., E.J. Carpenter, V.J.K. Iacono, 1983. Identification and enumeration of marine chroococcoid cyanobacteria by immunofluorescence. *Appl. Environ. Microbiol.* 46: 553-559.

Campbell L., L.P. Shapiro Haugen, L. Morris, 1989. Immunochemical approaches to the identification of the utraplankton: assets and limitations. In E. M. Cosper, V.M. Bricelj, E. J. Carpenter (eds.), *Novel phytoplankton blooms: causes and impacts of recurrent brown tides and other unusual blooms*, pp. 39-56. Berlin: Springer Verlag.

Dover C.L. Van., 1990. Biogeography of hydrothermal vents communities along seafloor spreading centers. *Trends Ecol. Evol.*, 5: 242-246

Euzet, L., J. F. Agnese and A. Lambert, 1989. The value of branchial Monogeneans for species identification of their hosts: a demonstration using a parasitological and genetic double blind study. *C. R. Acad. Sc.* 308: 385-388.

Frontier S., 1985. Diversity and structure in aquatic ecosystems. *Oceanog. Mar. Biol. Ann. Rev.* 23: 253-312.

Gentil F. and J.-C. Dauvin, 1988. Is it possible to estimate the total species number of a macrobenthic community? Application to several soft-bottom communities from the English Channel. *Vie Milieu* 38: 207-212.

Gibbs, S.P., 1990. The evolution of algal chloroplasts. In W. Weissener, D.G. Robinson and R.C. Stan (eds), pp. 145-157. New York: Springer Verlag.

Giovannoni S.J., T.B. Britschgi, C.L. Moyer and K.G. Field, 1990. Genetic diversity in Sargasso Sea bacterioplankton. *Nature* 345: 60-63.

Glynn P.W. and W.H. de Weerdt, 1991. Elimination of two reef-building hydrocorals following the 1982-83 El Niño warming event. *Science* 253: 69-71.

Grassle J.F., 1991. Deep-sea benthic biodiversity. *BioScience*, 41: 464-469.

Grassle J.P. and J.F. Grassle, 1976. Sibling species in the marine pollution indicator *Capitella* (Polychaeta). *Science* 192: 567-569.

Grassle J.F. and J.P. Grassle, 1977. Temporal adaptations in sibling species of *Capitella*. In B. C. Coull (ed). *Ecology of marine benthos*. Columbia, S.C.: University Press.

Grassle J.F., N.J. Maciolek Blake, 1990. Are deep-sea communities resilient? In G.M. Woodwell (ed.), *The earth in transition*, pp. 385-393. Cambridge University Press, New-York.

Grassle J.F., P. Lasserre, A.D. McIntyre and G.C. Ray, 1991. Marine biodiversity and ecosystem function. *Biology International* Special issue 23, 19 pp.

Gray J.S., 1989. Effects of environmental stress on species rich assemblages. *Biol. J. Linnean Soc.* 37: 19-32.

Groot, C., R. E. Bailey, L. Margolis, and K. Cooke, 1989. Migratory patterns of sockeye salmon smolts in the straits of Georgia, British Columbia, as determined by analysis of parasite assemblages. *Can. J. Zool.* 67: 1679-1678.

Hawksworth D.L. and R.R. Colwell, 1992. Biodiversity amongst microorganisms and its relevance. *Biology International* No 24: 11-15.

Hatcher B.G., R.E. Johannes and R.E. Robertson, 1989. Review of research relevant to the conservation of shallow tropical marine ecosystems. *Ocean. Mar. Biol. Ann. Rev.*, 27: 337-414.

Hempel G. 1991. Life in the Antarctic sea ice zone. *Polar Record* 27: 249-254.

Hoeh W.R., K.H. Blakley and W. Brown, 1991. Heteroplasmy suggests limited biparental inhteritance of *Mytilus* mitochondrial DNA. *Science* 251: 1488-1490.

Horner R.A., 1985. *Sea ice biota*. Boca Raton: CRC Press.

Huntley M.E., M.D.G. Lopez, and D.M. Karl, 1991. Top predators in the Southern ocean: a major leak in the biological carbon pump. *Science* 253: 64-66.

Hurlbert E.M., 1971. The nonconcept of species diversity: a critique and alternative parameters. *Ecology* 52: 577-586.

Jackson J.B.C., 1972. The ecology of molluscs of *Thalassia* communities, Jamaica, West Indies. II. Molluscan population variability along an environmental stress gradient. *Mar. Biol.* 14: 304-337.

Jackson J.B.C., 1991. Adaptation and diversity of reef corals. *BioScience* 41: 475-482.

Kinne O., 1971. Salinity - invertebrates. In O. Kinne (ed.) *Marine ecology* (vol. 1), *Environmental factors*, (vol. 2), pp. 821-995. Glasgow: Maclehose.

Knight I.T., S. Shults, C.W. Kaspar and R.R. Colwell, 1990. Direct detection of *Salmonella* spp. in estuaries by using a DNA probe. *Applied and environmental Microbiol.*, 56: 1059-1066.

Koehn R.K. and B.L. Bayne, 1989. Towards a physiological and genetical understanding of the energetics of the stress response. *Biol. J. Linnean Soc.* 37: 157-171.

Kohlmeyer J. and B. Volkmann-Kohlmeyer, 1991. Illustrated key to the filamentous higher marine fungi. *Botanica Marina* 34: 1-61.

Lalli C.M. (ed), 1990. *Enclosed experimental marine ecosystems: a review and recommendations*. New York: Springer Verlag.

Lasserre P., 1976a. Metabolic activities of benthic microfauna and meiofauna. In I.N. McCave (ed.), *The Benthic Bounary Layer*, pp. 95-142. New York: Plenum Press.

Lasserre P., 1976b. Osmoregulatory responses to estuarine conditions: chronic osmotic stress and competition. In *Estuarine Processes,* vol. 1. *Uses, Stresses and Adaptation to the Estuary*, pp. 395-415. New York: Academic Press.

Lasserre P., 1991. Holistic perception of the oceans, and their living resources: a new chanllenge? *Biology International* 23: 14-17.

Lasserre P. and J.M. Martin, 1986. Biogeochemical processes at the land-sea boundary. Elsevier Oceanography Series, 43. Amsterdam: Elsevier.

Lee C.C. and T. Fenchel, 1972. Studies on ciliates associated with sea ice from Antarctica. II. Temperature responses and tolerances in ciliates from Antarctic, temperate and tropical habitats. *Arch. Protistenkd.* 114: 237-244.

Levinton J.S., 1982. *Marine ecology*. Englewood Cliffs, N. J.: Prentice-Hall.

Mayer A.G., 1914. The effect of temperature upon tropical marine animals. *Pap. Tortugas Lab., Carnegie Inst. Washington* 6: 1-24.

McDonald J.H. and R.K. Kochn 1988. The mussels *Mytilus galloprovincialis* and *M. trossulus* on the Pacific coasts of North America. *Mar. Biol.* 99: 111-118.

McIntyre A.D. and J.B. Pearce (eds). 1989. Biological effects of marine pollution and problems of monitoring. *Rapports et ProcÁs-verbaux des RÄunions du Conseil Permanent International pour lExploration de la Mer*, 179.

Moore H.B., 1972. Aspects of stress in the tropical marine environment. *Adv. Mar. Biol.* 10: 217-269.

Moore C.G. and B.J. Bett, 1989. The use of meiofauna in marine pollution impact assessment. *Zool. J. Linnean Soc.* 96: 263-280.

Ogden J.C. and E.H. Gladfeter (eds.), 1983. Coral reefs, seagrass bed and mangroves: their interactions in the coastal zone of the Caribbean. *UNESCO Rep. Mar. Sci.*, N° 23, UNESCO, Montevideo, 133 pp.

Pawlik J.R., C.A Butman and V.R. Starczak, 1991. Hydrodynamic facilitation of gragarious settlement of a reef-building tube worm. *Science* 251: 421-424.

Pearson T.R. and R. Rosenberg 1978. Macrobenthic succession in relation to organic enrichment and pollution of the marine environment. *Oceanogr. Mar. Biol. Annu. Rev.* 16: 229-311.

Proctor L.M. and J.A. Fuhrman 1990. Viral mortality of marine bacteria and cyanobacteria. *Nature* 343: 60-61

Ray G.C. 1991. Coastal-zone biodiversity patterns. *BioScience* 41: 490-498.

Ray G.C. and J.F. Grassle 1991. Marine biological diversity. *BioScience* 41: 453-457.

Remane A. and C. Schlieper 1971. *Biology of brackish water*. New York: Wiley Interscience.

Rex M.A., 1973. Deep-sea species diversity: decreased gastropod diversity at abyssal depths. *Science* 181: 1051-1053.

Rohde, K. 1993. *Ecology of marine parasites*. Second Edition. Wallingford, Oxon: CAB International.

Ryther J.H. and R.R.L. Guillard 1959. Enrichment experiments as a means for studying nutrients limiting phytoplankton production. *Deep Sea Res.* 6: 65-69.

Sanders H.L. 1968. Marine benthic diversity: a comparative study. *Am. Nat.* 102: 243-282.

Solbrig O. T. (ed.) 1991. *From genes to ecosystems: a research agenda for biodiversity*. Report of a IUBS-UNESCO Workshop, Harvard University. IUBS Monograph Series, 124 pp.

Southward A.J. 1980. The western English Channel - an inconstant ecosystem. *Nature* 285: 361-366.

Spight T.M., 1977. Diversity of shallow-water gastropod communities on temperate and tropical beaches. *Am. Nat.* 111: 1077-1097.

Spindler M. 1990. A comparison of Arctic and Antarctic sea ice and the effects of different properties on sea ice biota. In U. Bleil and J. Thiede (eds.), *Geological history of the polar oceans: Arctic versus Antarctic*, pp.173-186. Berlin: Kluver-Verlag.

Spoel S. van der, and A.C. Pierrot-Bults (ed.), 1979. *Zoogeography and diversity in plankton*. Utrecht: Bunge Scientific Publishers.

Spoel S. van der and R.P. Heyman, 1983. *A Comparative atlas of zooplankton*. Berlin: Springer Verlag.

Steele D. H., 1988. Latitudinal variations in body size and species diversity in marine decapod crustaceans of the continental shelf. *Int. Res. ges. Hydrobiol.* 73: 235-246.

Steele J. H., 1974. *The structure of marine ecosystems*. Cambridge, Ma.: Harvard University Press.

Steele J. H., 1985. Comparison of marine and terrestrial ecological system. *Nature* 313: 355-358.

Steele J.H. 1991. Marine functional diversity. *BioScience* 41: 470-474.

Stehli F.G., A.L. McAlester and C.E. Helsley 1967. Taxonomic diversity of recent bivalves and some implications for geology. *Geol. Soc. Amer. Bull.* 78: 455-466.

Straube W.L., M. O'Brien, K. Davis, and R.R. Colwell, 1990. Enzymatic profiles of 11 barophilic bacteria under in situ conditions: evidence for pressure modulation of phenotype. *Applied and Environ. Microbiol.* 56: 812-814.

Tamplin M. L., A. L. Gauzens, A. Huq, D. A. Sack, and R. R. Colwell, 1990. Attachement of *Vibrio cholerae* serogroup 01 to zooplankton and phytoplankton of Bangladesh water. *Appl. Environ. Microbiol.* 56: 1977-1980.

Tarazona J., H. Salzwedel and W. Arntz 1988. Positive effects of "El Niño" on macrozoobenthos inhabiting hypoxic areas of the Peruvian upwelling system. *Oecologia* (Berlin) 76: 184-190.

Tchernia P. 1978. *Océanographie Régionale. Description Physique des Océans et des Mers*. Paris: ENSTA.

Tilman F. 1977. Resource competition between planktonic algae: An experimental and theoretical approach. *Ecology* 58: 338-348.

Varvio S.L., R.K. Koehn, R. Vainola, 1988. Evolutionary genetics of the *Mytilus edulis* complex in the North Atlantic region. *Mar. Biol.* 98: 51-60.

Vaulot D., F. Partensky, J. Neveux, R.F. Mantoura and C.A. Llewellyn, 1990. Winter presence of prochlorophytes in surface waters of the north eastern Mediterranean Sea. *Limnol. Oceanogr.* 35: 1156-1164.

Wanntorp H.E., D.R. Brooks, R. Nilson, E. Nylin, F. Ronfquist, S.C. Stearns and N. Wedell, 1990. Phylogenetic approaches in ecology. *Oikos* 57: 119-132.

Warwick R.M. and Ruswahyuni, 1987. Comparative study of the structure of some tropical and temperate marine soft-bottom macrobenthic communities. *Mar. Biol.* 95: 641-649.

Wieser W. and F. Schiemer, 1977. The ecophysiology of some marine nematodes from Bermuda: seasonal aspects. *J. exp. mar. Biol. Ecol.* 26: 97-106.

12. The role of mammal biodiversity in the function of ecosystems

Vladimir Sokolov

Introduction

Quantitatively, mammals account for a small proportion of the biosphere of the earth or of particular ecosystems. The number of mammalian species is relatively small, about 4 thousand, while, for instance, the class of birds has 8 500 species and the class of fishes, about 28 thousand. Yet, mammals are characterized by an extreme diversity in terms of morphology, physiology, ecology and behavior. Similar diversity can, perhaps, only be encountered in some invertebrates (but not at the class level), while it does not occur in any other vertebrate classes.

For instance, the body mass (weight) in mammals can range from 1.5 g in pygmy shrew (*Suncus etruscus*) to 120 thousand kg in the blue whale (*Balaenoptera musculus*), the ratio being 1 to 80 million. On the contrary in birds, the weight of a humming-bird, which is the smallest bird, is about 2 g, while that of the world's largest bird, the ostrich, is 90 kg, the ratio being 1 to 45 thousand.

A classical example of the morphological diversity of mammals is found in the diversity of their limbs, including pentadactyl, plantigrade, semi-plantigrade, digigrade, with a reduction of digits to one, digging and rowing limbs, and even a flying limb. The physiology of reproduction of mammals is also characterized by a considerable diversity of viviparity of young, from developed ones, capable of following the mother several hours after birth to those laying eggs.

The level of development of the higher nervous system ranges in mammals from a very primitive one, close to that of reptiles as in many Insectivora, to the most highly organized, as in higher primates.

The high degree of organization and versatile diversity have helped mammals to distribute widely throughout the Earth - they are only absent in the central regions of the Antarctica - and occupy the most varied environments: terrestrial, subterranean, arboreous and aerial. No other vertebrates are so widely distributed.

Mammal Diversity

The diversity of mammals determines their great importance in ecosystems. Unfortunately, the role of mammals, particularly that of their diversity in the functioning of ecosystems is very little understood (we are not referring here the role of humans). The effect of mammals on other constituents of the ecosystems is mainly accomplished via nutrition (particularly so in herbivores) and through the effect on the soil in digging and subterranean mammals.

In cases of population outbreaks of some mammalian species, they can badly affect vegetation and the soil. So, for example, in the steppes of Mongolia during some years, the Brand's vole (*Microtus brandti*) breeds in astounding numbers. In thousands of hectares of land, this animal almost completely destroys the vegetation, and the entire soil surface ends up covered by piles of earth thrown up by the voles from their burrows. The steppe ecosystem completely changes its face.

In steppe country, both in natural and in managed areas, an increase in the diversity of species of herbivorous mammals leads to a fuller and more versatile utilization of plant production. Each species of herbivorous mammals has a specific set of forage plants, which in full-member communities of herbivores reduces competition among plants, and allows the regular use of the entire diversity of plants by the herbivores. When some individual species of herbivorous mammals predominates they utilize only a small portion of phytocenosis production. In fact, in the Ukrainian steppes the marmot (bobak) (*Marmota bobac*) populations even at their maximum density utilize no more than 12% of plant production. In the semidesert steppes of the Northern Cis-Caspian Region, the little souslik (*Citellus pygmaeus*) population at its limit density (80 individuals per hectare) used only 20% of plant yield. The same applies to domestic livestock. Each individual species only uses a certain part of the entire set of plants available. In the steppes of Northern Kazakhstan, only 64% of the entire vegetation was used when cows alone were grazed. By contrast, when cows were grazed together with sheep, 88% of the vegetation was used. Thus, the diversity of species composition of grazing mammals is one of the most important conditions of the exploitation of forage resources and of an increase in the ecosystems' secondary productivity. The same holds for the savannas of East Africa inhabited by zebras, antelopes and gazelles.

An important beneficial effect of the increase in species diversity of herbivorous mammals is provided by measures aimed at preventing the phytocenoses from being polluted by unpalatable plants. The latter predominate when rangelands are used for grazing a single or few animal species. A striking example is the vegetation of the astrakhan rangelands of Uzbekistan (Samarkand Region), where intensive grazing of sheep alone brought about heavy overgrowth of the ranges by the unpalatable species *Peganum garmala*. The same is observed in the Ethiopian Reserve Senkele Hartbeest Sanctuary. The absolute predominance in the area of a single species of ungulate hartbeest (*Alcelaphus buselaphus*) at a density of 5 thousand per 3 thousand hectares of the Reserve has resulted in the dominance of an unpalatable species *Penicetum schempery* in the vegetation. The accumulation of a large amount of litter from this species is accompanied by annual fires in the Reserve during the dry season of the year. On the whole, the preservation in such ecosystems of high total plant productivity is accompanied by a sharp decrease in forage productivity of phytocenoses and a drop of the carrying capacity of ranges in reserves for this species.

In some cases, an increased production of plant species unpalatable for domestic livestock is beneficial to valuable (rare) species of wild animals. Apparently, the abundance of saigas (*Saiga tatarica*) in the rangelands of Kazakhstan that are intensively grazed by domestic livestock, and particularly so in Kalmykia is due to sheep husbandry in these regions. Such species as *Bassia sedoides, Gypsophila paniculata,* and *Thlaspi arvense* become dominant. These species, unpalatable to sheep, flourish under conditions of anthropogenic development of rangelands. These conditions prove beneficial to saiga antelopes, which favor these plants.

A decrease in species diversity of the community of complete exclusion of certain species of herbivorous mammals is accompanied by profound changes in the phytocenoses and ecosystems.

Changes in the Environment

Lower species diversity and lower numbers of herbivores result in the accumulation of large amounts of litter, which is very detrimental to the entire ecosystem. In the steppes of the Ukraine in protected areas, the amount of accumulated dead organic matter is 90 times that of rangeland areas (55.4 against 0.6 centners per hectare). In the Kursk meadow steppes (the Central Chernozem Reserve) in the protected area 58-84 centners per hectare of dead organic matter accumulates, while in meadow steppes where the grass is mown, only 13-14 centners per hectare accumulates. Grass mowing simulates its consumption by the now-extinct wild horse tarpan (*Equus gmelini*), the saiga antelope (*Saiga tatarica*), the marmot (*Marmota bobac*), and the little souslik (*Citellus pygmaeus*). Dead organic matter disturbs the temperature and illumination conditions of the phytocenosis, since during the season of plant growth, the temperature of the active soil layer and that of the air are lower, and the input of solar radiation smaller. As a result, growth of the vegetation decelerates, and the onset of phenological phases is retarded. Thus, in the Streletsky Reserve (Lugansk Region), the onset of spring vegetation can be a month late, and marmots have little forage available during this important season when the winter hibernation is over.

In the final analysis, the formation of a dead plant layer makes the conditions for the resumption and growth of plants so bad, that bare patches of earth appear that are later overgrown with weedy plants (*Carduus hamulosus, Onopordum acanthium,* etc.). These unfavorable changes in the environment also affect the habitats of numerous animal species, which finally disappear from such ecosystems. In steppe reserves of the Ukraine such characteristic species as the bobak and little and spotted sousliks as well the great jerboa have disappeared.

Finally, dead organic remains bind a large number of mineral nutrients and exclude them from the biological cycle, thereby reducing the availability of essential mineral substances to plants. In the Kursk meadow steppe, about 50-160 kg/ha of mobile (that is, accessible to plants) forms of nitrogen and 400-950 kg/ha of mobile mineral substances, including such deficient ones as potassium and phosphorus, are lost in this way. The capacity of the biological cycle declines substantially. Normally large quantities of these fertilizers are applied on croplands. In rangelands almost the entire plant mass is processed by the animals in the course of metabolism, and nutrients, in the form accessible to plants (carbon dioxide, ammonia, urea, etc.) are returned to the environment. On the assumption that in rangelands herbivores alone consume 60% of above ground phytomass, 30-40% of protein nitrogen and mineral substances bound in the above ground phytomass are split by animals in the course of metabolism, to say nothing of undigested plant remains that are returned to the environment in forms that are accessible to plants.

Effect of Animal Activity on Ecosystem Structure

The mechanical activities of animals (such as digging in rodents and trampling in ungulates) play an important role in ecosystems. Removal of particular animal species can affects the structure of the ecosystem. So, for example, ungulates loosen the upper layers of the soil permitting seeds to be set in the soil, and keep the soil from being overgrown with mosses, lichens and algae. An exclusion of animal mechanical activity results in depression of higher (vascular) plants, ageing of the phytocenoses, and decline of their productivity.

In protected areas of the semidesert of Northern Cis-Caspian Region (Janybek Research Station) the denuded surface of the soil (between the plants) is overgrown with the cyanophyte alga *Nostoc*, which forms a dense film, which swells when wet and is impermeable to water. In rangeland plots this film is broken up by ungulate hooves. In protected plots this film covers 97% of the entire bare soil surface, in the rangeland, only 37% (Table 1). This film has a great effect on soil properties, primarily its water permeability. This explains why water permeability of the soil in the protected plot is almost 5 times lower compared with the rangeland plot (Table 1).

It also should be mentioned that drastic changes have occurred throughout the entire steppe zone of our Planet, caused by human activities. Biodiversity of animals, and in particular, mammals, has not only been reduced, but the majority of steppe species of plants and animals have disappeared. Steppes have been destroyed and turned into croplands. This process started hundreds of years ago, and it is just now being completed. Yet, this process has not been calamitous to the Earth's Biosphere, and is even beneficial to mankind, since the steppe area is one of the sources of cereal crops, which form the basis of the diet of people and domestic livestock.

Marine Mammals

The mammalian population of the coastal waters of the Commander Islands is presently not rich in terms of species. However, 250 years ago there dwelled a wonderful animal, the sea cow (*Hydrodamalis gigas*). This beast populated the waters off Bering and Medny Islands. The sea cow was discovered by Steller in 1741 (Steller was a member of the Bering expedition). These huge animals somewhat more than 7 meters long kept in shallow water throughout the year, particularly near the mouths of rivers and streams.

The sea cows were not mobile, and spent most of their time eating. They consumed various algae, mostly *Agarum* sp., *Nereocystis* sp., *Alaria fistulosa*, *Laminaria longipes*, and the sea-grass *Zoostera* sp. Where they foraged at least for a day, the rejected roots and stalks lay on the shore in piles.

Table 1. Overgrowing of the soil surface with algae and changes in water permeability of heavy loams in the Northern Cis-Caspian Region after grazing was discontinued (Abaturov, Pochvovedeniye, 1991, No 8).

Index	Range	Reserve
Area covered by mosses and lichens (%)	37±10.0	97±0.6
Water permeability of soil, mm/min	0.240±0.019	0.055±0.01

Each sea cow consumed large amounts of sea plants. Steller wrote that the intestines of these animals were always filled tightly with food. The stomach was surprisingly large: 180 cm long and 150 cm wide. It was so filled with sea plants that four adult robust men, with all their strength, could hardly move it and drag it with a rope. The number of sea cows were very high, and hence, their effect on underwater vegetation, and through it, the entire coastal ecosystem, was substantial.

Among mammals dwelling in the same places as the cow were the sea otter (*Enhydra lutris*) and the ringed seal (*Phoca vitulina*). The former mainly fed on sea urchins, then in decreasing order of importance, bivalve molluscs, octopuses, crabs and fishes. The main item of the diet of the ringed seal was fish. Occasionally, the minke whale (*Balaenoptera acutorostrata*) and some dolphins could appear in the vicinity of the islands.

Fourteen years after the discovery of the sea cow by Steller, that is, by 1754, this animal was totally exterminated by people on the Medny Island, and by 1768 on the Bering Island. No special studies of possible changes in the shoreline ecosystem of the Commander Islands as a result of the extinction of the sea cow have been made. Available evidence indicates that no calamitous, or, at least, very noticeable changes occurred.

The second most important mammalian species of this ecosystem, the sea otter, was also almost exterminated. Unregulated harvest of the sea otter reduced their population on the Medny Island from 10 to 15 thousand in the 18th century to about 350. In the 1950s the sea otter population on the Medny Island increased to 450-500 and in the 1960s it rose to roughly 1500. Today their number is considerably higher. An almost complete absence of sea otters on the Commander Islands for at least 50 years has not caused any great changes in the ecosystems. Thus, this sad human experiment, has apparently made no substantial effect in terms of loss of mammalian biodiversity on the functioning of the marine coastal ecosystem.

However, humankind has lost Stellar's sea cow not only as an interesting animal species but an undoubtedly valuable potential domestic animal. In his diary Steller wrote: "This animal is originally tame, and one doesn't have to tame it." Consequently, had the species not been eliminated, where marine vegetation is rich, such as in the harsh conditions of the North, we could have had sea cow farms. Now this will never come to pass.

Another tragic example of loss of biodiversity of mammals in marine ecosystems occurred due to ill-regulated harvest of cetaceans. True enough, no cetacean species has become extinct, but the populations of a number of species have so declined that one can regard these species as virtually non-existentent.

Effect of Loss of Marine Mammals on Trophic Chains

Hypothetically, one can visualize the situation that may arise following the extinction of whales. These sea giants feed on zooplankton, and the discontinuance of whale pressure should cause an outbreak of zooplankton. That, in turn, should result in reduction of the numbers of phytoplankton. Phytoplankton being the mayor oxygen producer in the biosphere, its decline may reduce the oxygen content in the Earth's atmosphere, which would be disastrous.

Yet the actual situation is different. The zooplankton underconsumed by big baleen whales was immediately eaten up by other consumers. The bulk of zooplankton was eaten by birds, seals, minke whales, and presumably, by fishes. As demonstrated by British researchers, improved trophic conditions brought about some striking physiological changes in crabeater seals, including earlier onset of sexual maturity, a reduction of the interval between births and an increase in the proportion of pregnant females in the populations. In fact, according to evidence obtained on the Southern Georgia Island, in the crabeater seal (*Lobodon carcinophagus*), during the last 25 years the age of sexual maturity dropped from 5 years to 2.5 years.

The result was a rapid rise in the numbers of crabeater seals, their annual increment reaching 7.5%. The avian population has also sharply increased. 90% of the avian biomass in Antarctica is accounted for by different species of penguins.

During the last 20 to 30 years the penguin populations increased from an annual reproductive rate of 2-3%, to 6-10%. Owing to the increase in the populations of these animals the zooplankton not consumed by the disappearing baleen whales was completely consumed by them. In fact, the overall diet of seals and birds increased from 100 to 260 million tons per year, and the upper trophic level of the Antarctic community, which was previously cetacean, has become penguin-seal.

The changes in marine ecosystems which undoubtedly took place in the Northern Hemisphere in connection with decline of whale populations are almost unknown. One of the most common whales in the past in Arctic waters was the bowhead whale (*Balaena mysticetus*). A pre-exploitation world population of some 50,000 or more bowheads whales must have made a considerable impact on the northern standing stocks of planktonic and epibentic invertebrates. It has been estimated that a single bowhead eats 870 kg (wet weight) of plankton per day during a 165-day feeding season each year. Not only did bowheads formerly consume enormous amounts of prey, but their collective feces must have "fertilized" the Arctic seas to an important degree. The remnant bowhead stocks, because of their relatively small size, probably have a much smaller impact on their environment.

On the whole, there are reasons to believe that there occurred no drastic, or even disastrous, changes in the Biosphere as a result of decline of the biodiversity of marine mammals. True enough, for a long time, that is, until commercial populations of these animals were restored, mankind has lost a valuable source of food protein, food and industrial fats, vitamins, etc.

Other Examples of Loss of Mammal Biodiversity

In Russia, by the beginning of this century, due to unregulated harvest two species yielding excellent pelts, namely the beaver (*Castor fiber*) of the order Rodentia and the sable (*Martes zibellina*) of the order Carnivora, became almost completely extinct. The beaver is a semi-aquatic animal, populating the banks of some small forest rivers and streams, which flow slowly. It feeds on various bank and aquatic plants. The sable is a dweller of the taiga, plain or mountain, particularly of the Korean pine taiga. This is an omnivorous animal, preferring voles and Korean pine nuts. No particular changes in the ecosystem function were recorded after these animals almost disappeared.

Some 10 to 15 years ago, as a result of huge nature-conservation work, the populations of these animals increased manifold, and they virtually recovered their original ranges. Again, no changes in the ecosystems were noted. However, no special studies were made to this effect. In some places where the beavers dammed small rivers, causing them to flood and form shallow (or occasionally, large) waterbodies, the area was transformed into a bog and the forest died. There, of course, the ecosystems changed.

Biodiversity Changes due to Introductions

Investigations of the effect of a particular group of organisms on the functioning of the ecosystems can be conducted not only when species diversity is stable or when it diminishes, but also when it increases. In the case of mammals examples are provided by the introduction and acclimatization of certain species. In the USSR, the most large-scale acclimatization of mammals involved the muskrat (*Ondatra zibethica*). As a result that animal has populated numerous waterbodies of Belorus, Ukrraine, European Russia, Siberia, Kazakhstan and Central Asia. Its population has become enormous. In the 1950s-1960s several millions of muskrats were taken annually. No noticeable changes in the bog, river or lake ecosystems populated by muskrats were recorded. In the some waterbodies the muskrat had a detrimental effect on the indigenous semi-aquatic species of Insectivora the desman (*Desmana moschata*), although no rigorous proof to that effect is available.

The same applies to the acclimatization in the forests of the European USSR of the raccoon dog (*Nyctereutes procyonoides*), which was imported from the Far East. According to some scientists and hunters the transplantation of the raccoon dog had detrimental effect on the populations of land-nesting birds, for instance, the black grouse and the partridge. It appears that no other notable changes have taken place in the ecosystems.

Conclusions

To conclude, I would once again like to emphasize our far from complete understanding of mammalian diversity in the functioning of ecosystems. Loss of mammalian (and not necessarily mammals only, but rather any organism) species in the ecosystems under the destructive effect of people can bring about an evolutionary deadlock. Not only for the given group alone, but also for mammals of other taxonomic groups related to the group in question cenotically. Example are found in rodents and carnivores or in ungulates and carnivores.

While at the species level, we have studied mammals fairly well, we do not have yet objective criteria for the identification of subspecies and, particularly, populations of mammals. And, it is exactly populations that are the units of species existence which are actually functioning in the ecosystem. Thus, mammalogists have a huge work to do, firstly, to make an inventory of subspecies and populations of mammals, and secondly, to investigate the role of mammals in the functioning of the ecosystems.

One possible method of investigation of biodiversity is through the establishment of research stations. It appears very promising to set up a number of research stations on the transect crossing Siberia from the North - from the tundra zone on the Taimyr Peninsula and roughly along the Yenisei River through the taiga zone southward (the Siberian transect). There are presently two to three biosphere reserves there, and we plan to base our studies primarily on them. Subsequently, we plan setting up one to two stations in Mongolia (in the valley of the Ubsu-Nur Lake such a station has already been established). We are negotiating with Indian colleagues about the possibility of continuing the Siberian transect to India.

13. The role of biodiversity in the functioning of savanna ecosystems

Ernesto Medina and Otto Huber

Introduction

Savannas are ecosystems dominated by a more or less continuous grass layer and distributed mainly within the tropical belt. They are found in environments characterized by a strong seasonality in rainfall, generally with soils of low fertility and submitted to variable herbivore pressure and intensity and frequency of fires (Hills 1965; Walter 1973; Sarmiento 1984; Cole 1986; Frost *et al.* 1986; Medina and Silva 1990). Because of the different criteria used to define savannas in different parts of the world, it is practically impossible to give precise figures about their real extension. Area estimations range from 21 to 24 x 10^6 km^2 (Hills 1965; Cole 1986). Savannas occupy large extensions of the terrestrial surface of Africa (65%), Australia (60%), and South America (45%) (Cole 1986).

Savanna environments are characterized by the regular (rainfall) or more or less unpredictable (fire) incidence of stressors (*senso* Lugo 1988) on a nutritional substrate which is generally limiting for plant productivity (Medina 1987) (Box 1). Under these conditions biodiversity is bound to be reduced compared to other tropical terrestrial ecosystems, such as tropical rainforests (Lugo 1988). The evolution of biological diversity and the maintenance of species richness are subjects highly debated in the ecological literature, particularly its regulation through biotic (competition) and abiotic (stressor) factors (Connell and Orias 1964; Grubb 1977; Huston 1979; Lugo 1988). It is not the purpose of this paper to discuss the concept within the framework of savanna environments. From the outset we agree with the general statement that biological diversity appears to be regulated by (a) the availability of resources, in the sense that environments with too low or too high availability of resources through time tend to be less diverse (Tilman 1986) and (b) the incidence and intensity of stressors, resulting in lower diversity in the two extremes of the intensity of stress factors, such as drought, salinity, temperature, and nutrient availability (Lugo 1988) (Box 2). Availability of resources and intensity of a given environmental stress can be conceived as complementary. Environments with high level of resource availability are not stressful, allowing development of strong biological interactions, while low resource availability constitutes a stressful condition for the amount of energy that the organisms have to invest in capturing the limiting resource(s). Resource availability and stress intensity have opposite effects on biodiversity because the main impact of a certain stress on a plant community is to affect the relative competitive ability of the co-existing species. The most competitive species are those showing higher growth and productivity rates. Experimentally it has been shown that for a certain composition of strong and weak competitors, an increase in the intensity of a particular stress has a more pronounced deleterious effect on the stronger competitor (Bornkamm 1963). If the same mechanism operates under natural conditions, then it may open the site for the establishment of other species with lower competitive ability but less demanding for a given limiting resource.

In this paper we address the following questions: (1) how diverse are savanna ecosystems in terms of species and life forms; (2) is this diversity associated with differentiation in metabolism and ecophysiological properties of the species involved? (3) which is the ecological function of diversity in savanna environments?

Box 1. Tropical savannas are stress driven ecosystems

Regular drought

Rainfall seasonality with interannual variations in amount of water, duration of the rainny season (Rainfall Evaporation), and the beginning of the rainy season

General low soil nutrient status

 Low availability P, K and Ca. Nitrogen pool depending on the productivity of specific savanna vegetation and frequency of fire. Seasonal pulses of organic matter minerlization at the beginning of the rainy season.

Seasonal but unpredictable recurrence of fireaffects mineralization of above ground organic matter, and regrowth vigour of both herbaceous and woody plants. Too late or too early fires after the onset of drought are deleterious for plant performance (water availability for new growth by grasses, and elimination of newly produced leaves in trees).

Seasonal selective herbivory

Main impact during the growing season, resulting in variable reduction of productivity and reproductive effort.

To answer these questions we will have to restrict ourselves to the vegetational aspects in areas where we have more ecological experience, namely the savannas of South America, although relevant findings from other savanna areas of the world will be used when appropriate.

General Characteristics of Biological Diversity in Savannas

Savannas and similar tropical herbaceous ecosystems are usually considered to be less diverse and more uniform than forests or other ligneous ecosystems of the tropics. Certainly, considering only the total number of species, typical seasonal savannas rank far behind the forests. Yet if we analyze the various aspects of the wider range of savanna

Box 2. Relationships between biodiversity, resource availability, and the structure of stress and competition driven ecosystems. The variation from the left to the right hand side characteristics are not linear.

BIODIVERSITY

LOW   HIGH   LOW

BIOLOGICAL SYSTEM

(No. of species, energy and nutrient flows, interactions)

Stress Dominated   Competition dominated

main energy expenditure
for:

| nutrient acquisition, drought temperature, salinity, anoxia and photoin-hinbition resistance | maximize growth rates, competitive exclusion, symbiosis, niche regeneration |

LOW   HIGH

Resource availability

parameters: soil fertility, water supply (amount, periodicity), temperature, light
processes: mineralization, meteorization, and nutrient uptake; photosynthesis, transpiration, bionmass and nutrient allocation

ecosystems in the world, a more complex and at the same time more realistic view will be obtained.

Whereas the large scale geographical distribution of savannas in the tropics appears to be primarily correlated with the presence of a pronounced bi-seasonal rainfall regime (i.e. alternating dry and wet seasons), their regional distributional pattern is clearly more strongly influenced by local edaphic and biotic conditions. Thus, a mosaic of different savanna types occupies in some cases wide and continuous extensions with a relatively homogeneous climatic regime, such as northern and central South America, central and western Africa, southern India, and northern Australia. On the other hand, isolated savannas occurring in small patches interspersed in large forested regions, such as in the Amazon basin or in parts of south east Asia, are mainly the result of peculiar edaphic or historical situations.Typical South American savannas are characterized by having a floristically rich herb layer mixed with a variable proportion of ligneous elements. Generally, the predominant life form of the ligneous component is used as differential criterion for the definition of a particular savanna type. Thus, shrub savannas and tree savannas are clearly distinguished from pure, open savannas (such as the Brazilian gradient ranging from "campos limpos" [grassy savannas] to "campos sujos" [shrub savannas], "campos cerrados" [tree savannas] and "cerradao" [savanna woodlands or sclerophyllous forests]) (Goodland and Pollard 1973; Sarmiento 1983)

A similar situation can be observed in the African savannas, but in addition there the herb layer shows various degrees of differentiation into "tall grass savannas" and "low grass savannas" probably in response to the selective impact produced by the much higher density and diversity of large herbivores (Menault et al. 1985).

Floristics

Contrary to tropical forest ecosystems, where the constant predominance of certain plant families is rather the exception, savannas are invariably dominated in their herbaceous layer by only two large families, namely the grasses (Poaceae) and the sedges (Cyperaceae). In most savannas of the world, these two families account not only for the largest number of species, but they also constitute the main biomass of the ecosystem. For instance, in a well developed neotropical savanna, grasses number between 30 and 60 species, whereas there are usually from 15 to 30 sedge species. Third in floristic importance are the legumes (Fabaceae) which often reach twenty or more species in a given savanna type. Other important herb families are Asteraceae, Rubiaceae, Euphorbiaceae, Labiatae, Convolvulaceae, Scrophulariaceae, Gentianaceae, and Sterculiaceae (Table 1).

Species richness of neotropical savannas shows a wide range, both in the herbaceous and in the ligneous components. According to Sarmiento (1983), the number of grass species ranges between nine and more than 200, that of the non-grassy herb and subshrub species between 87 and 330, and that of the woody species between 15 and 249. It must be kept in mind however, that detailed floristic inventories are still missing from large savanna areas, thus rendering the above mentioned figures purely indicative.

Almost all South American savannas contain at least some ligneous plants, although occasionally these may be low subshrubs hidden within the herbaceous layerand therefore not noticeable at first glance. Such small subligneous forbs belong to various families, such as the Leguminosae (*Eriosema, Desmodium, Clitoria*), Asteraceae (*Pectis, Ichthyothere*), Amaranthaceae (*Gomphrena*) and Euphorbiaceae (*Phyllanthus*).

Table 1. **A.** Distribution of families, genera and species in representative samples of savanna in Brazil's Cerrado do Triangulo Mineiro, corresponding to about 2300 ha (Goodland and Ferri 1979).

Families	Genera	Species	Genera	No. spp.
Asteraceae	36	69	Cassia	16
Gramineae	31	73	Vernonia	14
Papilionoideae	25	52	Paspalum	11
Rubiaceae	17	30	Vochysia	11
Palmae	13	22	Hyptis	9
Euphorbiaceae	10	14	Eriosema	9
Caesalpinoideae	8	30	Panicum	9
Cyperaceae	8	24	Tecoma	8
Bignoniaceae	8	22	Annona	7
Malpphigiaceae	8	14	Aspidiosperma	7
Apocynaceae	7	15	Bulbostylis	6
Melastomataceae	7	11	Bauhinia	6
Annonaceae	5	11	Axonopus	6
Labiatae	5	13	Jacaranda	6
Mimosoideae	5	12	Rhynchospora	5
Vochysiaceae	3	16	Erythroxylon	5
			Andropogon	5
Total	**196**	**428**	Borreria	5
			Psidium	5
			Byrsonima	5
			Baccharis	5

B. Main families and genera (with more than 5 species) occurring in the savannas of central Venezuela, inventoried in the 250 ha Calabozo Biological Station, Guárico State (Aristiguieta 1966).

Families	Genera	Species	Genera	No. spp.
Gramineae	16	35	Cassia	8
Papilionoideae	10	25	Polygala	8
Asteraceae	8	12	Axonopus	6
Cyperaceae	6	12	Paspalum	6
Euphorbiaceae	4	10	Sida	6
Malvaceae	3	9	Andropogon	5
Caesalpinoideae	1	8	Euphorbia	5
Polygalaceae	1	8	Ipomea	5
Rubiaceae	4	8	Aristida	3
Labiatae	1	6	Desmodium	3
Other (23)	30	32	Phaseolus	
Total	**84**	**165**		

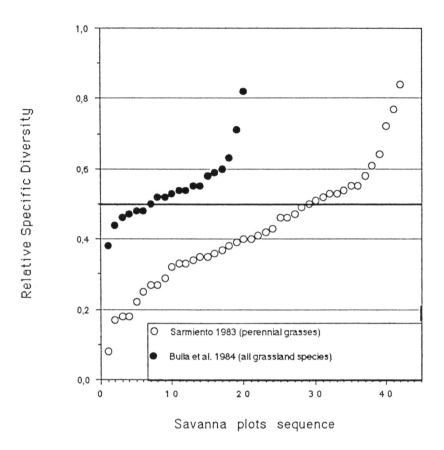

Figure 1. Relative diversity of savanna sites in Venezuela measured for the coexisting perennial grass species (Simpson index, Sarmiento 1983) and all the species of the herbaceous layer (Shannon-Weaver index, Bulla *et al.* 1984)

In the case of shrub savannas the ligneous compartment consists of a variable number of species belonging to a large group of families. Four families are particularly important in the neotropical shrub savannas: Dilleniaceae, with the almost exclusively savanna treelet *Curatella americana*; Malphigiaceae, represented by the large genus *Byrsonima*; Proteaceae (*Rupala*), and the palm family with numerous genera and species. Furthermore, each savanna region of South America has an additional set of typical woody savanna species such as *Salvertia convallariodora* (Vochysiaceae) and *Hancornia speciosa* (Apocynaceae) in the central Brazilian cerrados; *Antonia ovata* (Loganiaceae) in the northern Brazilian Rio Branco savannas and *Heteropteris laurifolia* (Malphigiaceae) in the Bolivian savannas. If the ligneous compartment of a savanna is formed by larger, usually evenly spaced trees, the number of arboreal species tend to be lower than in the case of shrub savannas. Often, only a few or even one tree species predominates, as for example in the peculiar Venezuelan tree savannas with *Platycarpum orinocense* (Rubiaceae) or with *Caraipa llanorum* (Guttiferae). Only in the case of "tree islands" occurring in savanna landscapes, the number of arboreal species augments rapidly, reaching a maximum in such savanna woodlands as the Brazilian "cerradao" or the Surinamese "savannen-bos."

Analyses of species diversity using conventional indices (Magurran 1988) provides two apparently contradictory pictures. Relative diversity of perennial grasses in a large set of neotropical savannas indicates that those with intermediate evenness

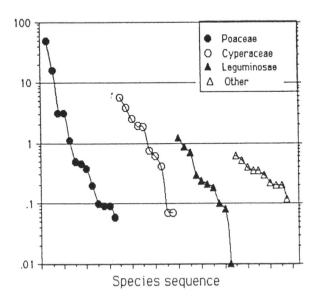

Figure 2. Dominance of species within the most frequent families found in savanna sites of the eastern Venezuelan savannas (Data from Bulla *et al.* 1984)

values predominate (0.42-0.45), although a significant number of savanna with lower indices are found. In a smaller number of savanna types, but including all species in the herbaceous layer, Bulla *et al.* reported even higher evenness values for a set of savanna plots in eastern Venezuela (Fig. 1). However in most data sets a strong dominance of a few grass species is observed (e.g. *Trachypogon plumosus* and *Axonopus canescens*) showing percentages of relative cover well above 10%. In each of the main families of these savannas there was a strong dominance of a few species (Fig. 2). In spite of a high tendency for species dominance, diversity is maintained by the occurrence of a number of relatively rare species.

Life-forms

Well developed savanna ecosystems with a continuous herb layer and a consistent shrub layer contain almost the entire spectrum of life forms in the classical sense of Raunkiaer, such as phanerophytes, chameophytes, hemicryptophytes, geophytes, and therophytes. However, as Sarmiento and Monasterio (1983) point out, these life form categories are not well suited for describing the more complicated situations commonly

148

Table 2. General life-forms in South American savannas. In each group the family containing the most important species is included.

	ANNUAL	PERENNIAL				
	Herbaceous [non-woody therophytes]	Herbaceous [non-woody hemicryptophytes, geophytes chamaephytes, vines, epiphytes]	Suffruticose [subligenous hemicryptophytes and chamaephytes]	Fruticose [ligneous shrubs chamaephytes, phanerophytes]	Arboreal [phanerophytes]	Lianas
GRAMINOID	Poaceae, Cyperac, Xyridac	Poaceae, Cyperac, Xyridac				
BROAD-LEAVED	Melastomatac, Polygalac, Rubiac	Melastomatac, Asterac, Rubiac	Lamiac, Asterac, Melastomatac	Dilleniac, Malphigiac, Fabac	Fabac, Bignoniac Proteac	Bignoniac Dilleniac
GROUND ROSETTE	Euphorbiac, Eriocaulac					
ROSETTE		Asterac, Morac, Caryophyllac				
EPHEMERAL	Lentibulariac					
MEGAPHYLLOUS		Heliconiac, Marantac				
VINES		Convolvulac, Fabac				
EPIPHYTES		Bromeliac, Orchidac				
PARASITES		Laurac, Convolvulac		Loranthac, Morac		

observed in tropical plant communities. Dominant life forms in savannas are those of perennial species, and within them two groups contain the largest number of species: the herbaceous graminoid and broad-leaved species and the sub-ligneous and ligneous hemicryptophytes (Table 2). (for a detailed discussion of life-forms in savannas see Sarmiento and Monasterio 1983).

Naturally, in species-rich savannas a higher diversity of life-forms is found than in savannas with a lower species number. Furthermore, among many of the cited life forms more specialized growth forms can be distinguished, For instance, the graminoid life form presents two major growth forms: the tussock or bunch grass growth form, and the solitary growth form. The former consists of a dense tuft of up to one meter or more ion diameter of perennial grasses or sedges, all arising from a common rootstock. This is by far the predominant growth form in the herbaceous layer of most savannas such as the *Trachypogon plumosus* (Poaceae) or *Scleria cyperina* (Cyperaceae) in many neotropical savannas. The solitary growth form, on the other hand is more commonly found in annual grass or sedge species, although some perennial grasses may also exhibit this feature. Sometimes solitary perennial grasses may even show a trailing habit as for example *Echinolaena inflexa*, a frequent grass species in many South American savannas.

Amongst the rosette life forms, various growth forms occur, such as ground rosettes, in which all leaves are condensed at the soil surface (e.g. many Eriocaulaceae) and the caulirosette growth form, in which the leaf rosette is concentrated at the apex of the vegetative axis more or less distant from the soil surface. This second growth form is found in certain sedges (*Bulbostylis paradoxa, B. lanata*) common in South American savannas on nutrient poor substrates.

Functional Significance of Life Forms

The grass-tree contrast in savanna ecosystems has long been recognized as a peculiar feature. There are numerous reports in the literature that the tree and herbaceous component in savannas are exploiting water and nutrient resources from different soil layer (Walter 1973; Walker and Noy-Meier 1982; Medina 1982; Goldstein and Sarmiento 1987). Neotropical savanna trees do not appear to be limited by water supply, apparently because they are able to exploit soil layers under the influence of the water table, while grasses show strong seasonal variations in water potential and a large degree of osmotic adjustment (Goldstein and Sarmiento 1987). However it has been proposed that both density and height of the tree layer in neotropical savannas is heavily dependent on the soil water availability during the dry season (characteristically associated with the depth of the water table, Medina and Silva 1990). From a dynamic point of view, the establishment of tree seedlings will depend on the occurrence of wet years, in order for them to withstand the competition from perennial grasses. Soil fertility also has been proposed as the determinant of the variation of the tree/grass ratio, particularly in the physiognomic sequence from savanna woodland (or sclerophyllous forest called "cerradao" in Brazil) to the open shrubby grassland ("campo sujo") described by Goodland and Pollard (1973) in the Triangulo Mineiro in Brazil. The reduction in the number of tree stems and species per hectare observed has been attributed to the reduction in nutrient availability (Fig. 3) and the increased Aluminium saturation of the exchange complex. In some cases these nutritional gradients might be

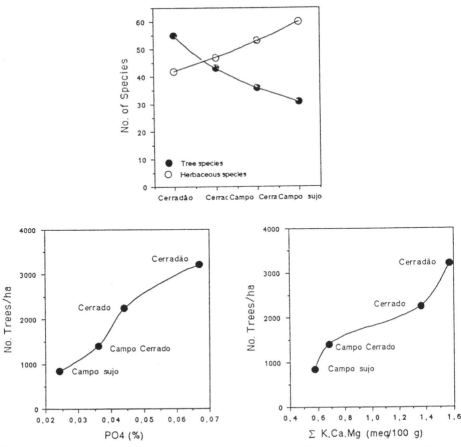

Figure 3. Variation in the number of trees per hectare in relation to the nutritional status of the upper soil layer along the vegetation sequence sclerophyllous forest (Cerradao), savanna woodland (Campo cerrado), wooded savanna (campo sujo), measured in the Triangulo Mineiro, Minas Gerais, Brasil (from data of Goodland and Ferri 1979).

the result and not the cause of vegetation development. This is especially so in the case when only the chemical composition of the upper soil layers is taken into account (Medina and Silva 1990). Moreover, differences in the density of the tree layer might be the result of variations in the frequency and intensity of the fire regime. Fire suppression for more than twenty years in the Biological Station of Calabozo, Venezuela, lead to a significant increase in the diversity of both the tree and the herbaceous layer (Fig. 4) and in an absolute increase in the number of stems per hectare of both savanna and forest tree species (Fariñas and San José 1987).

Much of the resistance of savanna ecosystems to environmental stresses such as drought, herbivory, and fire is associated with the development of underground organs. Most savanna ecosystems are characterized by large below ground/above ground ratios (Sarmiento 1983). This allocation of biomass to underground organs represents the conservation of energy and nutrients allowing the recovery of the grass,

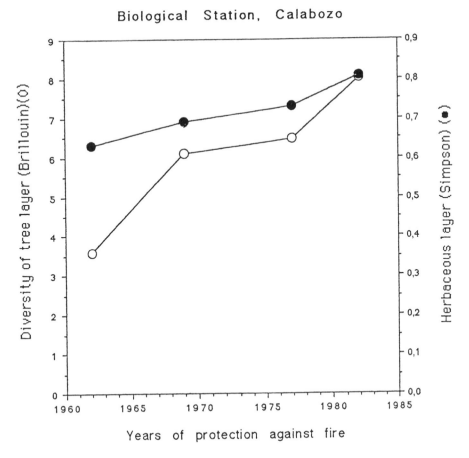

Figure 4. Variation in diversity of the tree and herbaceous layer in a 3 Ha savanna plot protected from fire since 1961 at the Biological Station in Calabozo, Venezuela (data from Fariñas and San José 1987).

shrub, and tree layers after regular, or even catastrophic events of drought, herbivory and fire.

Physiological Diversity and the Efficiency of Resource Utilization

Photosynthetic types, water-use efficiency and productivity

Savannas are found under a relatively large range of rainfall regimes, and on soils of widely different texture. Different combinations of these two factors can circumscribe regimes that vary from relatively dry to relatively wet. Therefore we find savannas all the way from semi-arid regions in Africa and Australia, to the border areas which are annually flooded because of a combination of high rainfall and heavy soils. This range of variation has allowed the classification of savannas into dry, intermediate, wet and swampy (Johnson and Tothill 1986), or slightly seasonal, seasonal, and hyper-seasonal savannas identified by Sarmiento in Tropical South America (Sarmiento and

Monasterio 1975). The analysis of the graminoid layer of savannas under different rainfall regimes has lead to a number of generalizations related to the taxonomic and metabolic differentiation of the grass family. Johnson and Tothill (1986) concluded that a few tribes of grasses have strong dominance in savanna environments. Those are the Paniceae (e.g. *Axonopus, Digitaria, Panicum, Paspalum, Setaria*), Andropogonae (*Andropogon, Elionurus, Gymnopogon, Heteropogon, Hyparrhenia, Imperata, Themeda, Trachypogon*), and Chlorideae (*Chloris, Microchloa*). Less important but frequent are the Oryzae (*Leersia, Luziola*) and the Arundinellae (*Arundinella*). The Andropogonae dominate the core of savanna ecosystems while the semi-arid savannas are dominated by representatives of the Chloridoid, Aristoid, and Eragrostoid tribes. The proportion of Panicoid increases strongly towards wetter savannas, and the Oryzoids appear dominating in the swampy savannas, or esteros of South America. However, the Panicoid have the largest ecological range, and this tribe appears heavily represented in all typical and wetter savannas. This pattern emerges clearly in gradients of dry savannas on sandy soils to flooded savannas frequently found in central Venezuela in the flooded savannas of Apure state in Venezuela, and the Gran Pantanal of Brazil (Table 3).

Table 3. Florula of flooded, wet and dry savannas in the southern Guárico State distributed by tribes (from Suchas 1984)

Savanna type:	Flooded	wet/sandy wet	dry/sandy dry
Andropogoneae	4	6	6
Paniceae	19	16	8
Eragrostideae	1	1	2
Aristideae		1	3
Chlorideae		1	1
Oryzeae	2		
Sporobolae			1

The implications of the variation in dominance of different grass tribes from arid and semi-arid savannas to wet and swampy (hyperseasonal) savannas lies in the differentiation of the photosynthetic metabolism. The herbaceous layer of semi-arid and seasonal humid savannas are absolutely dominated by C4 grasses (C4 acids being the first product of photosynthetic CO_2 fixation), while in swampy savannas frequently C3 grasses (phosphoglyceric acid being the first product of photosynthetic CO_2 fixation) reach the absolute dominance level (Medina *et al.* 1976a), or share the niche with C4 grasses of similar growth habit (Medina *et al.* 1976b). Perhaps the only notable exception to this distribution is *Echinolaena inflexa*, a C3 grass reported as dominant in shrubby savannas in Brazil (Goodland and Ferri 1979) or as a shade plant in the Cerradao by Klink and Joly (1989).

Within the C4 photosynthetic type further subdivisions can be made according to the first stable acid synthesized after CO2 fixation in the mesophyll, as malate formers, and aspartate formers, or according to the decarboxylation pathway of the respective C4 acids within the bundle sheath cells, mediated by NADP-malic enzyme (NADP-ME) in the malate formers and NAD-malic enzyme (NAD-ME) or PEP-carboxykinase (PCK) in the aspartate-formers (Gutierrez *et al.* 1974). It has been shown that aspartate formers predominate in dry savannas, while malate formers are the dominant photosynthetic type in humid savannas growing in distrophic soils (Ellis *et al.* 1980; Hatterley 1983). Within the aspartate formers is the NAD-ME which appears to predominate in areas of lowest rainfall (Ellis *et. al* 1980). These patterns of distribution are probably associated with water use efficiency higher in C4 than in C3 grasses, and with different nutritional requirements presumed to be higher in C3 species and among the aspartate formers in the C4 species.

The predominance of the C4 pathway of photosynthesis within the single most important family in savanna environments suggests that the higher water use efficiency and productivity associated with this photosynthetic pathway strongly affects success in a highly seasonal climate. Although a similar pattern of distribution of photosynthetic pathways can be detected within the Cyperaceae (C4 types in seasonal and dry savannas, C3 types predominating in hyperseasonal savannas, Medina unpublished), the great majority of dicot families in the savanna belong to the C3 type. Only within the genera *Euphorbia* (Euphorbiaceae) and *Pectis* (Asteraceae) have C4 species been reported. All leguminosae, the most characteristic plant group of tropical savannas besides the grasses, belong to the C3 photosynthetic type. When grown under optimal conditions legume species have always lower photosynthetic capacity than C4 grasses (Ludlow and Wilson 1971). The conclusion is that water stress is not the only determinant driving force structuring the herbaceous layer of savanna ecosystems.

Phenological spread as a means of resource partition

Sarmiento (1983) showed that the grass species in a large set of sampled savanna plots in Venezuela had distinct phenological behavior. They were classified as precocious, early, intermediate and late species according to their period of flowering and maximum growth. It was shown that in savanna plots with a set of six coexisting perennial grass species the periods of exponential growth of early, intermediate, and late species did not overlap, and that the intermediate and late growers accumulated larger above ground biomass than early growers (sarmiento and Monasterio 1983; Silva and Ataroff 1985). The phenological spread can be associated to certain patterns of biomass allocation in these grasses. The below ground to above ground biomass ratios decrease from early growers to intermediate, and to late growers (Silva 1987). Early growers probably have higher water use efficiency than late growers (Goldstein and Sarmiento 1987) and are able to exploit the early pulse of nutrient supply resulting from the mineralization of organic matter at the beginning of the rainy season. However, they do not reach the high levels of organic matter production attained by late growers. Phenological spread has also implications for the variability of the grass layer in tropical savannas. The full set of phenological behaviors can be found in humid tropical savannas with relatively short dry seasons. In this vegetation the delay in the onset of rain result in decreasing production of the early growers, while a shorter duration of the rainy season will affect production and reproduction of the late species (Medina and Silva 1990).

Physiological differentiation in nutrient-poor soils

Tropical savannas are well known by their low soil nutritional level (Medina 1987). Sarmiento (1990) indicates that the majority of neotropical savannas are distrophic or hyperdistrophic, as characterized by a sum of exchangeable bases ranging from 1-5 and less than 5 meq/100 g soil, respectively. Under these conditions the vegetation is expected to be very sensitive to changes in nutrient availability, and the nutritional requirements for normal growth of the co-occuring species should in general be lower compared to those of species of other grassland ecosystems growing on eutrophic soils.

The nutritional factor may contribute to the patterns of grass-sedges dominance observed throughout the savanna regions of the world, because in general these two monocot families have lower nutritional requirements compared to dicots, particularly the Leguminosae. In natural seasonal savannas in Venezuela all the legumes sampled appeared to be deficient in phosphorus and potassium, while the nitrogen reached similar values to those expected from cultivated forage legumes (Medina and Bilbao 1991). By contrast, the nutritional levels of the above ground biomass of grasses at the peak of their development in natural savannas appears to be only slightly below the level expected for cultivated forage grasses (Medina 1987).

A general pattern appears to be that grasses are mostly limited by the supply of nitrogen, while dicots, particularly legumes, are strongly limited by the availability of phosphorus, and the exchangeable bases level of the soils (Medina 1987, Medina and Silva 1990). Experiments conducted in savanna areas of Africa, Australia and South America allow the conclusion that increasing the supply of nitrogen, or of nitrogen and phosphorus together, increases significantly the organic matter production of natural grass swards. The growth of legumes, on the other hand, is strongly stimulated by increases in the phosphorus supply or by amendments to the soil pH through the addition of calcium salts. The increased growth is associated with higher phosphorus uptake, leading to improvements in photosynthesis and higher nodulation rates. Patchiness in nutrient availability in savanna soils would account for differences in the composition of the herbaceous layer. That is, conditions resulting in higher soil fertility, such as higher pH (lower aluminium mobility), lower phosphorus fixation after the process of mineralization of organic matter, and higher sum of bases, would favor legumes over grasses and sedges in the herbaceous layer. Patchiness therefore could account for much of the diversity observed in different savannas in the world.

Within each taxonomic group there are differences in the requirements of the limiting nutrients. Long term fertilization trials have allowed the detection of dramatic differences in growth responses of C4 grass species belonging to the aspartate- and malate-forming groups (Donaldson et al. 1984; O'Connors 1985). In an experiment conducted at Towoomba, South Africa it was shown that the species belonging to the Andropogonae tribe, particularly *Cymbopogon plurinodis, Heteropogon contortus*, and *Themeda triandra*, were reduced in percent cover at higher fertilization levels with nitrogen and phosphorous, or nitrogen alone, while the contrary occurred with species of the Paniceae such as Panicum maximum and the Eragrosteae such as *Eragrostis rigidior* (Fig. 5). Interestingly, all the species increasing in percent cover are aspartate formers, while those decreasing are malate formers. This appears to be consistent with the hypothesis that the aspartate formers have higher nutritional requirements, particularly nitrogen than the malate formers and explains the widespread dominance of the latter in seasonal savannas with distrophic soils (Huntley 1982).

Rhizopheric associations and symbiosis

The predominance of soils with intrinsically low nutrient availability has apparently given the ecological background for the development of a number of symbiotic or quasi-symbiotic associations in tropical savannas. The associations are: a) legume-*Rhizobia* symbiosis; b) mycorrhizal symbiosis, and c) the development of rhizosphere-associated microflora capable of improving nutrient supply to the plant through solubilization of nutrients or direct nitrogen fixation.

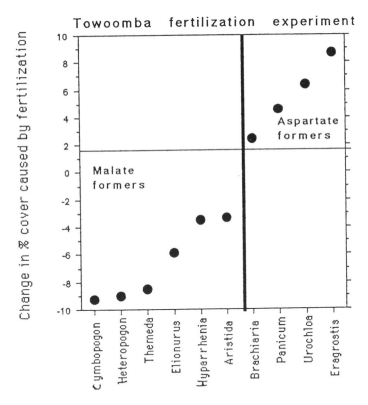

Figure 5. Variations in species dominance in long-term nitrogen fertilization trials conducted at Towoomba, South Africa. Data depicted correspond to the period 1949-1962 (Data from O'Connor 1985).

A large number of legume species of neotropical savannas, particularly within the Papilionoideae, have been shown to be nodulated in the field (Barrios and González 1971). The activity of these nodules has not been established, and a limitation to active nitrogen fixation might be expected due to the pronounced phosphorus deficiency in savanna plants. That this fixation might be significant, however, has been shown by a few measurements of the natural abundance of the heavier nitrogen isotope (^{15}N) made in leaves of legumes in Venezuela and West Africa. Several species within genera of the Papilionoideae, such as *Eriosema, Galactia,* and *Tephrosia* (Medina and Bilbao 1991, Abbadie *et al.* in press), have ^{15}N values well below zero, as expected for nitrogen-fixing species.

Legumes used as forage in tropical grasslands have been shown to be very responsive to phosphorus supply, and this supply can be provided in soils with low phosphorus availability by the formation of endo-mycorrhizal symbiosis (Saif 1986). Savanna legumes form vesicular-arbuscular mycorrhiza under natural conditions (Medina and Bilbao 1991). The importance of this symbiosis to the phosphorus supply has been shown quite dramatically under experimental conditions (among others Crush 1974; Arias *et al.* 1991) (Fig. 6). Most interesting is that the nodulation with *Rhizobia* is enhanced significantly probably as a result of the improved phosphorus supply to the plant mediated by the mycorrhizal symbiosis.

Rhizospheric associations with microorganisms of the *Azospirillum* type have been shown to be widespread in tropical savannas, and it has been reported that they are responsible for a significant amount of nitrogen supply to savanna soils (Balandreau 1976, Giller and Day 1985). However the evidence is not yet conclusive. If the rates reported from several sources are reliable, the rhizospheric fixation of nitrogen may constitute a significant contribution to the nitrogen balance of savannas. This field of research is of the utmost importance to fully understand the nutrient utilization by savanna grasses under natural conditions.

A clearer role in the nitrogen balance of savanna soils corresponds to the blue-green algae. Although it is still necessary to measure their actual nitrogen fixation at a large scale and over gradients of soil types and climates, the data available make it likely that these algae are fixing nitrogen in the order of kg/ha, that is, in amounts large enough to partially offset the nitrogen losses caused by annual grass burning (Medina 1987).

Conclusions

The previous analysis of the taxonomical, morphological, and physiological differentiation of the savanna vegetation shows that the role of biodiversity in ecosystem function resides mainly in the maintenance of system structure and the efficiency of resources utilization. The following characteristics can be uniquely associated with savanna vegetation, either as separate categories or as integrated properties:

a) Grasses with the C4 photosynthetic pathway dominate both the physiognomy and the production of organic matter in all seasonal savannas of the world. Within the grasses an additional differentiation of observed regarding the dominance of aspartate formers in drier savannas, and malate formers in humid, seasonal savannas with soils of low nutrient availability. The relationships between photosynthetic pathways and nutritional requirements have not been experimentally established, but long-term field experiments show clearly that fertilization with nitrogen favors the production of organic matter of aspartate-formers and reduces the competitivity of malate formers in grass swards where both types coexist.

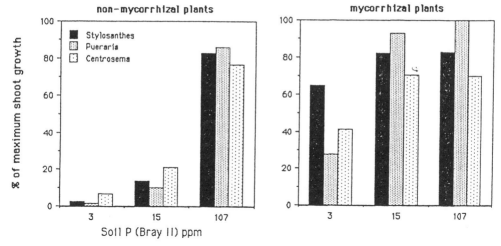

Figure 6. Differential response of tropical legumes to increasing availability of P in the soil, with and without mycorrhizal innoculation (from Arias *et al.* 1991).

b) The typical savanna structure includes the coexistence of trees and grasses. The competition for resource acquisition, particularly water, between these two contrasting life-forms is reduced by the differences in root distribution in the soil. Establishment of tree seedlings, however, might be limited to wet years, to allow the root of the tree seedling to cross the soil layer dominated by grass roots.

c) All savannas are characterized by the predominance of perennial species with strong development of underground organs, the root/shoot ration being generally above one. This biomass allocation pattern is very well related to the tolerance of savanna vegetation to stresses such as drought, fire, and herbivory.

d) Diversity of developmental patterns of savanna species, and the differentiation in nutritional requirements allows a segregation in space and time of the demand for resources. Grasses in general appear to be more dependent on nitrogen supply, while legumes and other dicots, are more limited by phosphorus supply.

e) The fact that savanna environments are nutrient-limited emphasized the significance of symbiosis for nutrient acquisition. Legume-*Rhizobium*-mycorrhizal symbiosis, and rhizospheric associations with free-living nitrogen-fixing bacteria, are very widespread, but their actual activity in nutrient supply at the ecosystem level has been measured only in a few cases. The activity of root symbioses, together with the ubiquitous presence of blue-green algae, provide the basis for establishing nutrient budgets in savannas, particularly the recovery of nitrogen losses induced by extensive burning.

References

Abbadie, L., A. Mariotti and J.-C. Menault. 1992. Independence of savanna grasses from soil organic matter for their nitrogen supply. *Ecology* 73: 608-613.

Arias, I., I. Koomen, J.C. Dodd, R.P. White and D.S. Hayman. 1991. Growth responses of mycorrhizal and non-mycorrhizal tropical forage species to different levels of soil phosphate. *Plant and Soil* 132:253-260.

Aristiguieta, L. 1966. Flórula de la estación biológica de los llanos. *Boletín de la Sociedad Venezolana de Ciencias Naturales*.

Balandreau, J. 1976. Fixation rhizospherique de l'azote (C_2H_2) en savanne de Lamto. *Revue d'Ecologie et Biologie du Sol* 13:529-544.

Barrios, S. and V. González. 1971. Rhizobial symbiosis in Venezuelan savannas. *Plant and Soil* 34:707-719.

Bornkamm, R. 1963. Erscheinungen der Konkurrenz zwischen höheren Pflanzen und ihre begriffliche Fassung. *Berichte des Geobotanischen Instituts Rübel* (Zürich) 34:87-197.

Bulla, L., P. Sánchez, C. Silvio, A. Maldonado, R. De Sola and A. Lira. 1984. Ecosistema sabana. In: *Bases para el diseño de medidas de mitigación y control de las cuencas hidrográficas de los rios Caris y Pao (Edo, Anzoátegui)*. Vol. 1, pp. 36-125. Caracas: Instituto de Zoología Tropical, Fac. de Ciencias, Universidad Central de Venezuela.

Cole, M.M. 1986. *The savannas: biogeography and geobotany*. London: Academic Press.

Connell, J.H. and E. Orias. 1964. The ecological regulation of species diversity. *Am. Nat.* 98:399-414.

Crush, J.R. 1974. Plant growth responses to vesicular-arbuscular mycorrhiza. VII. Growth and nodulation of some herbage legumes. *New Phytologist* 73:743-752.

Donaldson, C.H., G. Rootman and D. Grossman. 1984. Long term nitrogen and phosphorus application to the veld. *J. of the Grassland Soc. of South Africa* 1:27-32.

Ellis, R. P., J. C. Vogel, and A. Fuls. 1980. Photosynthetic pathways and the geographical distribution of grasses in South West Africa/Namibia. *South African J. 76: 307-314.*

Fariñas, M. and J. J. San José. 1987. Efectos de la supresión del fuego y el pastoreo sobre la composición de una sabana de *Trachypogon* en los llanos del Orinoco. In J. J. San José and R. Montes (eds.), *La capacidad bioproductiva de sabanas*, pp. 513-545. Caracas: Centro Internacional de Ecologia Tropical, Unesco-IVIC.

Frost, P., E. Medina, J. C. Menault, O. T. Solbrig, M. Swift, and B. Walker. 1986. Responses of savannas to stress and disturbance. *Biology International*, Special Issue No. 10. 82 pp.

Giller, K. E. and J. M. Day. 1985. Nitrogen fixation in the rhizosphere. In A. H. Fitter (ed.), *Ecological interactions in soils*, pp. 127-147. Oxford: Blackwell Scientific Publications.

Goldstein G. and G. Sarmiento. 1987. Water relations of trees and grasses and their consequences for the structure of savanna vegetation. In B. H. Walker (ed.), *Determinants of savannas*, pp. 13-38. Oxford, IRL Press.

Goodland, R. and M. G. Ferri. 1979. *Ecologia do cerrado*. Sao Paulo: Ed. da Universidade de Sao Paulo.

Goodland, R. and R. Pollard. 1973. The Brazilian Cerrado vegetation: a fertility gradient. *J. of Ecology* 61: 229-234.

Grubb, P. J. 1977. The maintenance of species-richness in plant communities: the importance of the regeneration niche. *Biological Reviews* 52: 107-145.

Gutierrez, M., V. E. Green and G. E. Edwards. 1974. Biochemical and cytological relationships in C4 plants. *Planta* (Berl.) 119: 279-300.

Hatterley, P. W. 1984. Characterization of C4 type leaf anatomy in grasses (Poaceae). Mesophyll:Bundle sheath area ratios. *Annals of Botany* 53: 163-179.

Hills, T. L. 1965. Savannas: a review of a major research problem in tropical geography. *Canadian Geographer* 9: 216-228.

Huntley, B. J. 1982. Southern African savannas. In B. J. Huntley and B. H. Walker (eds.), *Ecology of tropical savannas*, pp. 101-119. Berlin: Springer Verlag.

Huston, M. 1979. A general hypothesis of species diversity. *Am. Nat.* 113: 81-101.

Johnson, R. W. and J. C. Tothill. 1985. Definition and broad geographical outline of savanna lands. In J. C. Tothill and J. Mott (eds.) *Ecology and management of the world's savannas*, pp. 1-13. Canberra: Australian Academy of Science.

Klink, C. A. and C. A. Joly. 1989. Identification and distribution of C3 and C4 grasses in open and shaded habitats in Sao Paulo State, Brazil. *Biotropica* 21: 30-34.

Ludlow, M. and G. L. Wilson. 1971. Photosynthesis of tropical pastures. I. Illuminance, CO_2 concentration, leaf temperature and leaf-air vapour pressure difference. *Australian J. of Biological Sciences* 24: 449-470.

Lugo, A. 1988. Diversity of tropical species: questions that elude answers. *Biology International* Special Issue 19, 37 pp.

Magurran, A. E. 1988. *Ecological diversity and its measurement.* Princeton, N.J.: Princeton University Press.

Medina, E. 1987. Nutrient requirements, conservation and cycles in the herbaceous layer. In B. H. Walker (ed.), *Determinants of Savannas*, pp. 39-65. Oxford, IRL Press.

Medina, E. and B. Bilbao. 1991. Significance of nutrient relations and symbiosis for the competitive interaction between grasses and legumes in tropical savannas.

Medina, E. and J. F. Silva. 1990. Savannas of northern South America: a steady state regulated by water-fire interactions on a background of low nutrient availability. *J. of Biogeography* 17: 403-413.

Medina, E., T. de Bifano and M. Delgado. 1976a. Diferenciación fotosintética en plantas superiores. *Interciencia* 1: 96-104.

Medina, E., T. de Bifano and M. Delgado. 1976b. *Paspalum repens* Berg., a truly aquatic C4 plant. *Acta Científica Venezolana* 27: 258-260.

Medina, E., A. Mendoza and R. Montes. 1978. Nutrient balance and organic matter production of the *Trachypogon* savannas of Venezuela. *Tropical Agriculture* 55: 243-253.

Menault, J. C., R. Barbault, P. Lavelle, and M. Lepage. 1985. African savannas: biological systems of humification and mineralization. In J. C. Tothill and J. Mott (eds.) *Ecology and Management of the World's Savannas*, pp. 14-33. Canberra: Australian Academy of Science.

O'Connor, T. G. 1985. A synthesis of field experiments concerning the grass layer in the savanna regions of southern Africa. *South African National Scientific Programmes Report* No. 114. Pretoria: FRD, CSIR.

Saif, S. R. 1986. Vesicular-arbuscular mycorhizae in tropical forage species as influenced by season, soil texture, fertilizers, host species, and ecotypes. *Angewandete Botanik* 60: 125-139.

San José, J. J. and J. Garcia-Miragaya. 1981. Factores ecológicos operacionales en la producción de materia orgánica en las sabanas de *Trachypogon. Boletin de la Sociedad de Ciencias Naturales* 36: 347-374.

Sarmiento, G. 1983a. Patterns of specific and phenological diversity in the grass community of the Venezuelan tropical savannas. *J. of Biogeography* 10: 373-391.

Sarmiento, G. 1983b. The savannas of tropical America. In F. Bourliere (ed.) *Tropical Savannas*, pp. 245-288. Amsterdam: Elsevier.

Sarmiento, G. 1990. Ecologia comparada de ecosistemas de sabanas en América del Sur. In G. Sarmiento (ed.), *Las sabanas americanas: aspectos de su biogeografia, ecologia y utilización*, pp. 15-56. Caracas: Fondo Editorial de Acta Científica Venezolana.

Sarmiento, G. and M. Monasterio. 1975. A critical consideration of the environmental conditions associated with the occurrence of savanna ecosystems in Tropical America. In F. B. Golley and E. Medina (eds.), *Tropical Ecological Systems*, pp. 223-250. Berlin: Springer-Verlag.

Sarmiento, G. and M. Monasterio. 1983. Life forms and phenology. In F. Bourliere (ed.) *Tropical Savannas*, pp. 79-108. Amsterdam: Elsevier.

Silva, J. 1987. Responses of savannas to stress and disturbance: species dynamics. In B. H. Walker (ed.), *Determinants of Savannas*, pp. 141-146. Oxford, IRL Press.

Silva, J. and M. Ataroff. 1985. Phenology, seed crop and germination of coexisting grass species from a tropical savanna in Western Venezuela. *Acta Oecologica, Oecologia Plantarum* 6: 41-51.

Tilman, D. 1986. Resources, competition and the dynamics of plant communities. In M. J. Crowley (ed.) *Plant ecology*, pp. 51-75. Oxford: Blackwell Scientific Publications.

Walker, B. H. and E. Noy-Meir. 1982. Aspects of stability and resilience of savanna ecosystems. In B. J. Huntley and B. H. Walker (eds.), *Ecology of Tropical Savannas*, pp. 577-590. Berlin: Springer Verlag.

Walter, H. 1973. Die Vegetation der Erde in ökp-physiologischer Betrachtung. Band 1. Die tropischen und subtropischen Zonen. Jena: VEB Gustav Fischer Verlag.

14. Global change, shifting ranges and biodiversity in plant ecosystems

Thomas Van der Hammen

Introduction

Present day natural biodiversity and biodiversity patterns on earth are the result of a very long history, in which speciation, evolution, migration, competition and extinction are the processes. These processes take place in and are influenced by the environment, which is more or less complex, and by environmental change. This means that past global environmental change resulting in shifting ranges, influenced and partly determined biodiversity. Thus, if we want to understand present day biodiversity and if we want to know what may happen because of natural or human induced global change, then the study of the geological past is most important. Continental drift by plate tectonics, the upheaval of mountain ranges, volcanic events, the appearance or disappearance of islands and island chains, the appearance and disappearance of physical barriers and climatic change are all factors that may be involved in the course of geological history. A special and very important type of climatic change is represented by the Quaternary ice ages: a sequence of 100,000 year glacial-interglacial cycles during the last 2,5 million years.

I will try to illustrate all this with the example of two areas: western Europe in the holarctic floral area, with at present a temperate climate and relatively species poor, and northern South America, in the Neotropics, with some of the species-richest ecosystems of the world.

The Holarctic Floral Area

In the earlier Tertiary, North America was still lying near to Eurasia, forming together a large Laurasian northern continent, separated from the southern continents by the Tethys sea. In the north of this continent a temperate flora developed; in the southern part a (sub)tropical flora. As time went on, there was a gradual cooling of the planet, and in the upper Tertiary the (sub)tropical flora migrated southward and the (warm-) temperate flora gradually covered larger areas towards the South.

In the Miocene and Pliocene, (warm) temperate forests covered large areas of N. America, Europe and Asia, with a woody flora relatively rich in species. At this time, but especially during the Pliocene (some 5 million years ago), cooler episodes interrupted the otherwise (warm) temperate climate. This culminates finally (since 2,5 million years) in a long sequence of glacial periods, hen much of the original forest vegetation had to migrate southward and was replaced by tundra vegetation (in the north), or steppes in the Mediterranean area. While migration southward of vegetation and species was relatively easy in parts of America and eastern Asia, where mountain ranges and valleys have a north-south direction, in Europe the migration was much more difficult or impossible, because the principal mountain chains block the way southward, while south of these chains the drier climate (steppe vegetation) made tree growth very difficult. The only glacial refugia for forest elements were stretches or patches of montane forest in the Balkan, pressed between alpine vegetation above and steppe vegetation below, and anyhow difficult to reach. All this resulted in Europe in successive phases of extinction of elements of the (warm)temperate flora, and after every cold phase or glacial, often a considerable number of species and genera had become extinct, thus resulting in a considerable impoverishment of the W. European forest ecosystems (Van der Hammen *et al.* 1971).

On the other hand, this depletion did not take place in the same measure SE Asia, or in eastern North America. The general cooling of the climate during the Neogene (Upper Tertiary) resulted also in the migration southward of the subtropical flora in both SE Asia and N. America, where some of them found refuge in the montane zones of the tropical mountains (even into the Northern Andes, as we will see, where they enriched the flora with trees of the curious amphi-pacific floral element).

The Neotropics

Let us now turn to the Neotropics. We will consider both the Amazonian and Andean biota and ecosystems. The area is the most species rich of the world (there may be some 90,000 species). According to Gentry (1982) approximately half of the neotropic species are of Amazonian origin and half of Andean origin. In view of the geologically relatively young age of the Andes, this is quite surprising. West of the northern Andes lies the Chocó area. Here again is an extraordinary case with rainfall between 5000 and 10,000 mm (one of the wettest areas of the world).

Gentry (1986a) suggests that "the Chocó is not only one of the most strongly differentiated regions of the Neotropics phytogeographically, but that it has both the perhaps most diverse plant communities in the world and extremely high levels of local as well as regional endemism. "He also shows that there is a positive relation between rainfall and species richness in rain forest, increasing up to 4000 m, and then stabilizing at some 250 sp./1000 m^2. The Amazonian rain forest has between 100 and 250 sp./1000 m^2. Although one may think that Amazonia is very homogeneous as to vegetation, being covered principally by tropical rain forests, there is considerable differentiation of forest vegetation, and many centers of endemism are recognized 26 centers on the basis of woody angiosperm families.

The forest refuge theory says that these probably were forest refuge areas during dry periods when part of the rain forest would have been replaced by savanna or dry forest types. Speciation could have taken place in these refugia and the cyclic successive separation of populations and renewed distribution possibilities could have been an important motor of speciation, diversity in general or simple species richness. The theory has been much debated and partial alternatives have been offered of which one is that of internal ecological differentiation (Gentry 1986c) and barriers, causing a high beta-diversity. There are many indications (e.g. ecomorphologically) that relatively dry climatic conditions existed in the area in the relatively recent geological past, but to define possible forest refugia and their occurrence in time these data are in general insufficient. Some relatively recent data are however, of considerable importance. We know of dated dry phases in the SE and SW part of Amazonia, when sand dunes occupied areas at present covered by rainforest; these are of Holocene age (that means that they occurred during certain intervals of the last 10,000 years).

Palynological data from Rondonia (Brazil) indicate replacement of rain forest by open grass savanna at one time in the geologically recent past, and a dated pollen sequence from Carajás (SE Amazonia) shows that rain forest was replaced by savanna, approximately 60,000, 40,000 and especially 20,000-10,000 years ago, corresponding to the coldest and at the same time driest phases of the last glacial period (Van der Hammen 1972; Absy and Van der Hammen 1976; Absy *et al.* 1991; Van der Hammen and Absy 1994). If we add to this the fact that in the Orinoco savannas of Colombia and Venezuela, just north of the rain forest area, large dune fields developed turning this area into a type of desert, at or shortly before 11,000 - 12,000 B.P., it will be clear that northern South America suffered very strong climatic changes, partly correlated with glacial phases in the Andes and in the temperate and cold areas of the world. Hence the probability that the forest refugia theory is (at least partly) right, has increased considerably. So it seems that in this case global climatic change combined with considerable ecological diversity contributed greatly to species richness and flor-istic diversity of rain forest ecosystems. If high rainfall, ecological diversity and global cyclic climatic change, together sufficiently explain the high Amazonian diversity, is as yet an unanswerable question. There is certainly a long Tertiary history of the vegetation of the neotropical lowlands in which even periods of local brackish water presence occur, and we know from palynological data of the gradual enrichment because of the appearance of families or genera, such as Bombacaceae and *Mauritia* during the Paleocene, *Alchornea* and Malpighiaceae during the upper Eocene, Asteraceae during the Miocene etc. and on the other hand of extinctions. In this respect, it is interesting that lower Miocene biodiversity (species richness) of the river plain environment of western Amazonia may have been twice that of the present one: resp. 280 and 140/0.1 ha (Van der Hammen and Absy, 1994). Plio-Pleistocene extinction may therefore have been a factor of considerable importance, and the high western Amazonia species richness (Duivenvoorden 1994) might be partly due to a relatively low extinction rate in the very large North West Amazonian refuge area that did suffer less dry climate intervals (Van der Hammen and Absy, 1994). What we know for sure is that, although there was some exchange via island areas, the neotropical flora developed during the Tertiary in relative isolation, being sepa-rated from North America; the connection between North and South America was established definitely when the isthmus of Panama came into being, some 5 million years ago. And this facilitated the migration of plants and animals in both directions. This was especially important for this flora and vegetation of the Andes.

Diversity of the Andes

The high endemism and diversity of the Chocó area should be related to the upheaval of the Andean mountain chain, that separated it definitely from the interior of the continent (the present day Amazonia) and must have caused a considerable increase in rainfall. In the tropical Andes a very rich flora and very diverse vegetation types developed in the course of time: as we mentioned above, about half of the species of the very rich Neotropical flora are of Andean origin.

The principal upheaval of the Andes took place in the Pliocene, from about 5 million to 3 million years ago. At the same time, some 5 million years ago, the Panama land-bridge (where before an island chain existed) came more definitely into existence. However, the history of the Andean flora and vegetation should be much longer. During different intervals of the Tertiary, the area of the present day Andes was very mobile. It consisted largely of tropical lowland where fluvio-lacustrine or estuarine sedimentation took place and where locally peat was formed in extensive tropical lowland marshes. But in this lowland area, compression and vertical tectonic movements along faults created local chains of hills. During the earlier Tertiary these might mostly not have exceeded altitudes of 1000 m, but during the Miocene locally hills or mountains up to 2000 m may have existed.

The continuous processes of erosion acted on these hills and mountains lowering them, and although we sometimes know how much the rocks of a certain massive had been uplifted over a certain time interval, the erosion acting simultaneously may have caused that the surface never rose that much geomorphologically. Anyhow, it seems reasonable indeed to accept that hills not higher than 1000 m may locally have existed in the area during the lower Tertiary, that hills or mountains up to 2000 m may locally have existed during the Miocene and that mountains up to 5000 (or locally even to 7000 m) came into existence during the Pliocene, that persisted during the entire Quaternary up to the present day. This means that subandean lower montane environments were created in isolated areas, surrounded by neotropical lowland vegetation, already in the Miocene, while Andean or upper montane environments and páramo environments were created during the Pliocene.

If this is true, it would mean that the very rich subandean flora evolved principally in some 15 million years, while the rich andean and high andean floras developed and built up their species richness in some 5 million years (Van der Hammen et al. 1973; Van der Hammen and Cleef 1983/84, 1986). Still, in hills up to 1000 m in the Lower Tertiary, the basis for the subandean flora may already have been laid. In this respect it is interesting to note that according to Gentry (1982, 1986b) a very high rate of endemism (and active speciation) exists in the andean foothills, specially in epiphytes, understory shrub and palmetto genera.

Present day phanerogram species richness of montane andean forests is negatively correlated with altitude (temperature) and positively with rainfall. When the foothill area is wet, the number of species per 500 m^2 may decrease from some 150 in the lower part to 100 around 2000 m altitude and some 70 near the altitudinal forest limit. On the slopes towards the relatively dry tropical interandean valleys, the number of phanerogram species per 500 m^2 is only about 50 in the lower part, increasing to values of 100 around 2000 m, then decreasing again towards the forest limit (Rangel 1993; Van der Hammen 1993). Bryophyte species richness per 500 m^2

on the same slopes rises from only a few species at the tropical dry level to more than 50 to 2500 m, and then decreases towards the forest limit to some 25 species. Liver-moss species richness is relatively higher in those montane zones that have a more contantly high relative air humidity (Van der Hammen 1993). Species richness in the montane Andes is clearly related to climate, the wet warm tropical climate giving optimal conditions, the colder and drier climates being less favourable. However, another factor that as yet is difficult to evaluate, may be the time that an altitudinal zone has existed and during which the flora of that zone evolved and differentiated; this might equally lead to lower species richness in the progressively higher areas.

Islands

During the gradual enrichment of the andean flora, the presence of mountain top or hilltop environments as relatively isolated "islands", being created, being joined or separated, or disappearing, may have been important, separating populations, ex-changing species, and with founder effect phenomena. When the Panama land-bridge had been formed, first subtropical or warm temperate elements of the Laurasian holarctic Tertiary flora that had moved southward during the upper Tertiary cooling of the climate, migrated into the (lower) montane zones of the Andes, enriching the biogeographical diversity of their vegetations with amphi-pacific elements, locally form-ing almost 10% of the genera (Van der Hammen and Cleef 1983; Van der Hammen 1989; *Trigonobalanus* is an impressive example). Elements from the south, like *Weinmannia*, may also have entered at that time. When the mountains became higher, an increasing number of genera entered the upper montane zone from the north (holarctic floral area and south Austral-antarctic floral area).

Immigration was most notorious in the open andean páramo, nowadays restricted to areas above 3300-3800 m, forming a considerable number of isolated "islands" above the andean forest zone. During glacial times, the páramo extended far down, locally to 2000 m, and many islands were then united to much greater ones, and could exchange species. During interglacial times, like nowadays, páramo-refuge areas were formed in restricted higher areas. This stimulated speciation in isolation, what together with the immigration and consequent evolution of founder species from else-where, led to the gradual increase of species richness of the paramo floras and ecosys-tems, and in increase of biogeographical diversity. Genera of the paramo flora are for approximately 50% of (neo)tropical (incl. andean) origin, while 50% are of holarctic or austral-antarctic origin (Van der Hammen and Cleef 1986; Cleef and Chaverri 1992). All these factors playing a role during the Pliocene and Quaternary, a very species rich and diverse flora came into being, in the high mountains from northern Peru to Costa Rica, with important centers of diversity in the Colombian Eastern Cordillera and the Cordillera de Merida (Cuatrecasas 1979). It is estimated that this paramo flora contains some 3000-4000 species of flowering plants,, with up to 60% endemics (Luteyn 1992). Comparison of the differences in diversity (species richness) between different major paramo areas may give us information about the influence of the different processes in time. There are some 260 genera in the paramo flora of the Cordillera Oriental of Colombia (Van der Hammen and Cleef 1986), but some 80 of these genera do not occur in the Ruiz/Tolima massive of the Colombian Central Cordillera (Salamanca 1991). The isolated Santa Marta páramo has 145 genera, Tatama massive in the Western Cordillera 114, and the Talamanca paramo (Costa Rica) 140 (Cleef and Chaverri 1992). It is of course clear that size and isolation plays a role here.

However, the difference between the Eastern and Central Cordillera is further il-lustrated by the species of the well known genus of Asteraceae with stem-rosettes: *Espeletia s.s.* (Cuatrecasas 1986)There are 36 species in the Cordillera Oriental, and only 2 in the Cordillera Central. The differences in diversity seem to be related with two facts. The Central Cordillera has a relatively little interrupted paramo zone, while that of the Eastern Cordillera is very much fragmented(refugia); on the other hand the Eastern Cordillera is a sedimentary area, while the Central Cordillera is strongly volcanic. So, while on one hand speciation is stimulated in refugia, the effect of periods of strong volcanic eruptions should lead to extinctions (Salamanca 1991). We may conclude therefore that species richness and in general biodiversity in the Andes and its ecosystems, seems to be strongly influenced or determined by their earlier and more recent geological past.

Global Change and Biodiversity: Lessons from History

We have seen in the foregoing that this is not only true for the Andes, but also for Amazonia and Western Europe. We will never be able to explain fully certain aspects of biodiversity, if we do not take into account this history. In a certain way you might say that biodiversity besides its spatial pattern has a fourth dimension time.

What can we learn from all this, if we want to know what may happen under the influence of present-day and near future global change? As long as there was continuous natural vegetation on earth, a next strong climatic change, like a next ice age, would certainly change again completely the face of the earth, resulting in extinction, migration and speciation. But the earth has suffered that many times already, and would recover. A warming of only 2° or 3° C would bring back warm interglacial conditions with some shifting ranges, but without harm.

However, the situation has changed in most parts of the earth. With the exception of some areas, fragmentation of the terrestrial natural vegetation cover has advanced so much, that often only small fragments of forests or other natural vegetation are left. These fragments are mostly too small to maintain their species richness on the long run, while the intermediate areas became barriers difficult to pass for many organisms. In that case, as history teaches us, extinctions will occur in the case of notorious climatic change. In the case of a human induced rise of temperature of 2° or 3° C, this problem of barriers might be limited to the northern and southern limits of species areas. However, other climatic change caused by such a rise of temperature, like changes in the rainfall pattern on earth may also create problems along rainfall gradients. Mostly, however, the effect will probably be limited as long as it does not exceed the given values. I think that for the future, however, we should take into account the possibility of larger natural or human induced global climatic change, that could have a disastrous effect on the remnants of natural terrestrial ecosystems on earth. This means the need of very large nature reservations in the tropics (taking into account centers of diversity and endemism) and large scale regeneration of natural vegetation in the most affected areas of the world, especially in the temperate areas. And for the entire earth: corridors of natural (eventually regenerated) vegetation between the reservations, along the temperature and rainfall gradients. For the moment, however, that other type of global change, the disappearance of natural vegetation, especially forests, because of direct destruction by humans, is much more dangerous for biodiversity. In the tropics especially, the situation in the near future may become disastrous for the rain forests (including montane forests) that have a far greater biodiversity than any other biome; probably half of the species of the world are concentrated on less than 10% of the surface of the earth (Myers 1986). Here, more than anywhere else, mass-extinctions will occur increasingly if nothing is done to halt the Destruction.

It is true that our knowledge of the causes of biodiversity is very limited and the possibilities to predict what exactly will happen to ecosystem functioning is completely insufficient. It is certainly necessary to intensify our scientific efforts. But, it is also clear that more definite answers cannot be expected in the next decade(s). If we wait for that moment, most natural ecosystems and many of their species will have disappeared and there will be no time nor conditions for the creation of new species.

Hence, although the definite answers to the questions of what is going to happen in the near future are not yet available, I think we can learn some lessons from history and our responsibility as scientists requires that we tell the international community clearly and loudly, what we think is going to happen. Because, if no far-reaching measures for conservation and regeneration are taken now, there will be very little to be studied in the near future, and our late answers will be of little importance. There is no time to be lost.

References

Absy M.L. and T. van der Hammen, 1976. Some paleoecological data from Rondonia, southern part of the Amazon Basin. *Acta Amazonica* 6: 293-299.

Absy, M.L., A.M. Cleef, M. Fournier, L.Martin, M. Servant, A. Sifeddine, M. Ferreira da Silva, F. Soubiés, K. Suguio, B. Turcq and T. van der Hammen, 1991. Mise en évidence de quatre phases d'ouverture de la forêt dense dans le sudest de l'-Amazonie au cours des 60.000 dernières années. Première comparaison avec d'autres regions tropicales. *C.R. Acad. Sci. Paris* 312 Serie II: 673678.

Baumann, F., 1988. *Geographische Verbreitung und Ökologie Südamerikanischer Hochgebirgspflanzen. Beitrag zur Rekonstruktion der quartären Vegetationsgeschichte der Anden.* Physische Geographie Vol. 28. 206 pp. Geographisches Institut der Universität, Zürich.

Cleef, A.M. and A. Chaverri, 1992. Phytogeography of the Páramo flora of the Cordillera de Talamanca, Costa Rica. In H. Balsler and J. L. Luteyn (eds). *Páramos: an Andean Ecosystem under Human Influence*, pp. 45-60. London: Academic Press.

Cuatrecasas, J. 1979. Comparación fitogeográfica de páramo entre varias Cordilleras. In M.L. Salgado-Labouriau, (ed.), *El Medio Ambiente Páramo*, pp. 89-99. Caracas.

Cuatrecasas, J. 1986. Speciation and radiation of the Espeletiinae in the Andes. In F. Vuilleumier and M. Monasterio (eds). *High Altitude Biogeography*, pp. 267-303. New York: Oxford Univ. Press.

Duivenvoorden, J.F., 1994. Rain forest plant diversity in the middle caqueta area, Colombia, Western Amazonia. *Biodiversity and Conservation*, 3, in press.

Gentry, A.H. 1982. Patterns of neotropical plant species diversity. *Evol. Biol.* 15: 1-84.

Gentry, A.H., 1986a. Species richness and floristic composition of Chocó region plant communities. *Caldasia* 15(71-75): 71-91.

Gentry, A.H. 1986b. Endemism in tropical versus temperate plant communities. In E. Soulé (ed.), *Conservation Biology*, pp.153-181. Sunderland, Mass.: Sinauer Ass.

Gentry, A.H. 1986c. Sumario de patrones fitogeográficos neotropicales y sus implicaciones para el desarrollo de la Amazonia. *Rev. Acad. Col. Cienc.* 16 (61): 101-116.

Kroonenberg, S.B., J.G.M. Bakker and A.M. Van der Wiel, 1990. Late Cenozoic uplift and paleogeography of the Colombian Andes: constraints on the development of high-andean biota. *Geologie en Mijnbouw* 69: 279-290.

Luteyn, J.L. 1992. Páramos, why study them? In: H. Balsler and J.L. Lufteyn (eds). *Páramos, an Andean Ecosystem under Human Influence*, pp. 1-14. London: Academic Press.

Myers, N. 1986. Tropical deforestation and a mega-extinction spasm. In E. Soulé (ed.). *Biological diversification in the tropics*, pp. 137-158. New York: Columbia Univ. Press.

Rangel, J.O., 1993. Diversidad y frecuencia de las familias, generos y especies de plantas vasculares en el transecto Parque los Nevados. In T. Van der Hammen and A.G. dos Santos (eds). *Studies on Tropical Andean Ecosystems* 4. Chapter 10. Berlin-Stuttgart: J. Cramer (Borntraeger).

Salamanca, S. 1991. *The vegetation of the páramo and its dynamics in the volcanic massif Ruiz-Tolima* (Cordillera Central, Colombia). Ph.D. Thesis, University of Amsterdam.

Van der Hammen, T. 1972. Changes in vegetation and climate in Amazon basin and surrounding areas during the Pleistocene. *Geol. Mijnbouw* 61: 641-643.

Van der Hammen, T. 1989. History of the montane forests of the Northern Andes. *Plant Systematics and Evolution* 162: 109-114. Berlin: Springer Verlag.

Van der Hammen, T. 1993. Zonal ecosystems of the West and East flanks of the Colombian Central Cordillera, Parque de los Nevados Transect. In T. van der Hammen and A.G. dos Santos (eds) *Studies on Andean Ecosystems 4*. Chapter 27. Berlin-Stuttgart: J. Cramer (Borntraeger).

Van der Hammen, T. and M.L. Absy, 1994. Amazonia during the latest glacial. *Paleogeogr. Paleoclimat. Paleoecol.*, May.

Van der Hammen, T. and A.M. Cleef, 1983. *Trigonobalanus* and the tropical amphi-pacific element in the North Andean forest. *Journ. Biogeogr.* 10: 437440.

Van der Hammen, T. and A.M. Cleef, 1983/84. Datos para la historia de la flora andina. *Rev. Chil. Hist. Nat.* 56: 97-107.

Van der Hammen, T. and A.M. Cleef, 1986. Development of the high Andean paramo flora and vegetation. In F. Vuilleumier and M. Monasterio (ed.). *High altitude tropical biogeography*, pp. 153-201. New York: Oxford Univ. Press.

Van der Hammen, T., J.H. Werna and H. van Dommelen, 1973. Palynological record of the upheaval of the Northern Andes: a study of the Pliocene and Lower Quaternary of the Colombian Eastern Cordillera and the early evolution of its High Andean biota. *Rev.Palaeobot.Palynol.* 16: 1-122.

Van der Hammen, T., T.A. Wijmstra and W.H. Zagwijn, 1971. The floral record of the Late Cenozoic of Europe. In K.K. Turekian (ed) *The Late Cenozoic Glacial Ages*, pp. 391-424. New Haven and London: Yale University Press.

15. Shifting ranges and biodiversity in animal ecosystems

Jan H. Stock

Introduction

Rather than treating all animal ecosystems all over the world, I will concentrate in this paper on the aquatic ecosystems of the Atlantic and surrounding areas. The atmosphere is heating up due to the greenhouse effect. Ice caps, especially the North Pole, are melting, with, as a result, a world wide rise of the sea level, estimated at some 10-20 cm in the 20th century and at 3 to 8 times the present rate in the next century. Large economic effects are threatening our society, such as flooding of low lying coastal areas and salinization of freshwater supplies and agricultural grounds. The purpose of this meeting is not so much to present a pessimistic forecasts, but discuses the influence of a globally changing climate on biodiversity.

As Professor Van der Hammen showed so clearly (chapter 14), present day biotas did not fall out of the blue sky, but evolved through geological time from predecessors. I have chosen to look at global climatic changes in the geological past, and extrapolate the data obtained to the near future of the 21st century.

In the past 65 million years (My), the earth has witnessed two large episodes of enormous change: at the boundary of the Mesozoic and the Tertiary eras (the M/T boundary, approximately 62 My ago) and that between the Oligocene and Miocene (the O/M boundary, c. 30 My ago). While some scientists believe that the M/T boundary, marked for instance by the extinction of the dinosaurs all over the world, was caused by a cataclysmic event, viz. the full hit on earth by some planetary object, we are pretty sure that the O/M boundary was, at least in and around the Atlantic Ocean, due to climatic changes of important magnitude. As compared with later climatic changes, such as the Ice Ages, the O/M events were really of much greater importance. The earth before the Miocene did not look at all like the place we know today. Both the distribution of landmasses over the globe, and, consequently, the influence of the sea on the land, took more or less their present shape during the early Miocene. It might be appropriate to glance briefly at these events, and their effects on biodiversity, in order to understand better the global change expected in the next decades.

Global Changes in the Oligocene/Miocene

The Atlantic Ocean did not exist in the earlier Mesozoic; this ocean started forming, as a huge fault opening between the continents of Europe and Africa on the one side and the Americas at the other, in the late Mesozoic (c. 120 My ago). This cleft was caused by continental drift, driving the continental blocks apart. The cleft started and deepened out first in what is now the South Atlantic, but the continental blocks remaining hinged in the north. In the neighborhood of Greenland, America and Europe were still in contact with one another in the Oligocene; the hinge broke open at the O/M boundary, thus some 30 My ago. At first the fault was shallow, allowing enormous amounts of Arctic surface water to enter into the Atlantic, and by a reverse current, Atlantic warm surface waters entered the Arctic. The temperature in the surface waters of the Atlantic lowered. For example, in the tropical western Atlantic (the West Indies), it dropped by at least 5 degrees (some authors believe 7-10 degrees), as can be deduced from the temperature-dependent ratio of oxygen isotopes in fossil remains. Therefore the West Indian waters became too cool for two tropical marine ecosystems: the coral reefs and the mangrove forests. Many reef and mangrove species migrated southward, and many found a climatically suitable refugium on the coasts of Brazil.

On the other hand, the warm Atlantic waters streaming north through the fault caused massive melting of the polar ice cap; the melting water raised the sea level all over the world by some 100 to 200 meters. Coastal areas were flooded, and the warm-climate forests growing in the coastal areas were destroyed by the rising sea water. The remains of the vegetation were carried into the sea, where it started rotting to such an extent that deeper waters of more than 200 m in depth of the Atlantic were rapidly depleted of oxygen, and in the end became anoxic. Of course the entire deep water fauna of those parts of the ocean where anoxia occurred, was killed off. Because the North Atlantic surface water cooled, less evaporation was the result. This in turn caused less precipitation, in particular in the Arctic, and the ice caps became smaller and smaller: the loss by melting was no longer compensated by fresh precipitation. The influence of the cooler and drier climate upon terrestrial ecosystems will be clear at once, by what happened in the ocean. At first the tropical reef and mangrove communities tried to migrate south; when, through changed sea currents (the reason for which will be given later), the temperature of the surface waters rose again, corals and mangroves re-invaded the empty Caribbean niches and established the tropical biocenoses we know today.

The Temperature Crisis

What was the difference before and after the temperature crisis? Before the crisis some 500 species of coral lived in the tropical Atlantic, after the crisis only some 50 species. Thus 90% went extinct. Mangroves (a tropical tree community growing in salty waters) were represented by 40-45 species, of which only 4 survived (again an extinction of some 90%). Less habitat specific animals were less severely touched. For example, among sea urchins the diversity dropped from 73 to 26 species, some 60% decline. Many of the survivors re-immigrating into the empty niches of the West Indies underwent rapid evolution in the short period of a couple of million years. This evolution made them different from species or genera surviving elsewhere in the world. Others stayed as they were, or developed only subspecific differences. Another geomorphological feature of the earth that changed drastically at the O/M boundary was the distribution of sea/land around the equator. During the whole Mesozoic and in the Tertiary up to the Oligocene, the Mediterranean Sea was much larger: a circum-global, tropical sea, called the Tethys Sea. In this sea, the sea currents were from east to west, likewise circumglobal. At the O/M boundary, landmasses rose up at two places, fragmenting the Tethys Sea:

(1) the isthmus of Panama, at first an open seaway, rose above the sea surface, closing off the connection between the Pacific and the Atlantic; and

(2) the connection of the primordial Mediterranean and the Indian Ocean was closed by dry land that formed in the Middle East.

The tropical, circumglobal sea current stopped. Marine animals and plants that were first identical all over the tropics of the Tethys Sea got isolated in smaller sea areas and started independent lines of evolution. In the Atlantic, the circum global sea-current could not find its way out into the Pacific, had to turn around in he Gulf of Mexico, and formed what we now know as the Gulf Stream. This stream influenced the climate of enormous areas on the coasts of North America and Europe.

A third change, already briefly mentioned earlier, is that terrestrial materials originating from the inundated coastal areas, reached the sea causing lack of oxygen in the deeper layers (below the 200 m line) of the ocean. The Tethyan deepwater fauna got extinct by the anoxia. The situation improved when the fault between Europe and America, near Greenland, deepened to attain a depth of 3 to 4 km. Through this deep channel, a cold underflow (some $2°$ C) of salty water, originating from freezing sea water at the Arctic ice cap, began to flow to the south, near the bottom of the ocean. Because this water was both cold and was formed near the surface, it was rich in oxygen. It started the re-ventilation of the Atlantic depths, and made life possible again. Before the O/M events, the deep waters were $20°$ or more, after the event the temperature dropped by some $18°$. A rapid evolution of shallow-water ancestors, mainly from polar origin, into the new deepwater habitat took place, giving rise to the highly diversified deep sea fauna as we know it today. From the fossil record, we know for instance that the entire deep water Ostracod fauna got extinct and the end of the Oligocene, but by the end of the Miocene, already an entirely new deep-water ostracod fauna had evolved.

Global Change in the Next Decades: Lessons from Oligocene/Miocene Events

What can we learn from these Oligocene/Miocene events, and how can we extrapolate this knowledge to the global change expected for the next decades?

(1) Drastic changes in the physico-chemical environment (such as the depletion of oxygen in the ocean's deep waters) drive complete faunas to extinction. When the physico-chemical conditions improve, the empty niches are rapidly filled by new species evolving from populations that survived ("rapid" in this context might be some 1 to 10 My).

Is an oxygen crisis in the oceans to be expected in the coming years? Yes, a mild crisis is to be expected, because the rise of the sea level, followed by erosion and transportation of organic material into the sea will cause increased rotting in the ocean. However, whereas the O/M rise of the sea level was between 100 and 200 meters, the present rise of less than half a meter to at most 1.6 m will make the oxygen depletion much less spectacular and probably of local significance, e.g. in the Gulf of Bengal where enormous quantities of terrestrial material will end up in the sea.

(2) Less drastic changes (like a gradual change of temperature) will cause a shift of the ranges of species, although the results will be nothing compared with the O/M events. First of all, a limited and very gradual heating of the atmosphere will not eliminate entire biocenoses; most species will just change their distribution boundaries or migrate to refugia (of course refugia can act as centers of rapid evolution, and thus be a positive agent in the biodiversity process). We witnessed after the Ice Ages that many cold-adapted species got to refugia in the high mountains in central and southern Europe, got extinct in the central and western European lowlands (which were too warm for them), but survived in northern Europe. As we know from both morphological and im-munogenetic studies, the isolated relict populations in the mountains differentiated in a relatively short time (100,000 yrs) into subspecies a gain in biodiversity. Frogs (*Rana temporaria*) and newts (*Triturus alpestris*) show for instance this phenomenon. Is the expected rise in temperature big enough to influence biodiversity in a negative way? In most cases: NO. The change in temperature should be very considerable (in a magnitude of 5°C or more) to result in dramatic changes in biodiversity on a global scale. One should bear in mind that the annual temperature cycle in most of Europe and North America, is of a magnitude of 40-50°C on land and of some 20°C in the sea, hence a small temperature increase will be experienced just as "noise" of the annnual tempera-ture cycle and will not influence the biodiversity. Only on a local scale, and only in a limited number of species, mostly living under very specialized conditions and/or in very restricted habitats, extinction will take place. Examples of such specialized animals are ground water organisms, that live in very small geographically restricted groundwater veins; when the salinity of the ground water rises they will not be able to go elsewhere.

I feel that the direct human threat, e.g. the land use, and the pollution and poisoning of rivers, sea and land, presents a much greater threat to biodiversity on global scale than the greenhouse effect. Many rare species, e.g. now living in nature reserves, especially in wetlands, will not be resistant to a drier and warmer climate, whereas suitable new habitats which can act as refugia, are not available, being already in use by man.

Finally, I may point out that some scientists suppose that drastic changes in climate, affecting the oceans (a strong rise of the sea level, inundation and destruction of near coast biotas, resulting in an anoxic, rotting layer in the sea) show a periodicity of some 32 My. The last anoxic event was 30 My ago, the next to last 62 My ago, and the antepenultimate 94 My ago. If this assumption is valid, the last anoxic phase being 30 My ago, we are heading within the next 2 My to another crisis in the world's biotas. This new crisis will be unrelated to the hothouse effect, and perhaps the rise of temperature we are supposed to witness nowadays is just the preamble to a secular event of much greater amplitude.

Acknowledgments

In this paper I have used the data of many scientific investigators, in many different fields (oceanography, paleontology, biology, geophysics). However, for the way I used (or misused) their results in the present paper, I alone will be responsible. Many references pertaining to this lecture can be found in another synthetic paper of mine (Stock 1990)

References

Stock, J. H. 1990. Insular groundwater biotas in the (sub)tropical Atlantic: a biogeographic synthesis. *Atti dei Convegni Lincei, Accademia Nazionale dei Lincei, Roma* 85: 695-713.

16. Life-history attributes and biodiversity. Scaling implications for global change

Dean L. Urban, Andrew J. Hansen, David O. Wallin, and Patrick N. Halpin

Introduction

Concern with the preservation and management of biological diversity (Soulè 1986) has been amplified by recent speculation about anthropogenic global change (Peters and Darling 1985, Graham 1988, Schneider 1989, Smith and Tirpak 1989). By global change we mean to encompass climatic change--the so-called greenhouse effect--as well as large-scale changes in land use and its associated environmental consequences.

This concern presents two challenges: first, to accelerate the basic process of scientific research toward practical applications, and moreover, to extend or extrapolate our current state of knowledge to scales which ecologists (at least) are not used to managing. These are daunting problems but pressing concerns; we need to move quickly to meet these challenges.

Mechanism and Constraining Context

Biodiversity can be explained in terms of an interplay between life-history mechanisms (the demographic processes of growth, recruitment, and mortality), and environmental constraints (resource availability, disturbance regimes). In a hierarchical perspective, life-history processes are lower-level mechanisms while environmental patterns in resource availability or other boundary conditions are higher-level constraints (Allen and Starr 1982, O'Neill *et al.* 1986, Urban *et al.* 1987). For example, bird population dynamics can be viewed as the consequence of the environmental constraints of food and nest site availability acting on the processes of habitat selection, breeding, dispersal, and mortality of individual birds; larger-scale constraints may include landscape pattern, regional disturbance regimes, and so on (Figure 1).

Scaling problems arise when we attempt to extrapolate information across scales while losing track of important mechanisms or ignoring variability in higher-level constraints. A simple example of such a scaling miscue would be to extrapolate population densities based on a small (ha) censused area to a much larger area (km^2), without correcting for the patchiness of available habitats at the larger scale. Under the specter of global change, these scaling issues could degrade our ability to make valid predictions of the possible consequences of global change on biodiversity.

Here we overview some lessons we have learned about ecological factors governing biodiversity, and the implications of these in scaling up to global concerns. We do this in four vignettes, beginning at a small scale (a forest stand), and proceeding to larger scales (landscape, subcontinental, and global). The vignettes focus initially on forest birds but we believe the lessons are general to conservation practice and biodiversity.

Four Vignettes

1. Microhabitat Pattern and Avian Community Diversity

Microhabitat diversity at the scale of the forest stand (ca. 1-10 ha) is a fundamental constraint on avian community diversity (Urban and Smith 1989). The ecological mechanisms underlying this reflect behavioral aspects of habitat selection by forest birds. James (1971) coined the term "niche gestalt" to refer to the characteristic suite of habitat features (chiefly vegetation structure) associated with the breeding territories of forest songbirds (Figure 2). This concept has underpinned a rich legacy of multivariate statistical studies of habitat relationships (Capen 1981, Verner *et al.* 1986). The variety of such habitat "niches" represented within a forest stand is related directly to the number and variety of birds supported by the stand (MacArthur and MacArthur 1961, MacArthur *et al.* 1962, Roth 1976). The natural pattern of forest development is such that microhabitat diversity tends to increase with time; or, at a given snapshot in time, diversity increases with area (MacArthur and Wilson 1967, Connor and McCoy 1979).

Urban and Smith (1989) used a forest simulation model to illustrate the importance of microhabitat pattern to forest bird communities. They simulated 750 yr of forest dynamics for a 9-ha stand, and summarized microhabitat diversity in terms of the relative abundance of understory versus overstory trees. Bird "species" were defined as random niches (ellipses) in this understory/overstory 2-space, and potential species abundance was estimated in terms the amount of forest habitat falling within the niche space of each species at each time step of the simulation. Simulated community-level patterns in species abundance and diversity corresponded well to patterns observed in real bird communities (Figure 3), which emphasized the fundamental importance of microhabitat pattern to bird community structure.

Phenomenologically, this mechanistic relationship allows one to predict avian diversity as a simple function of stand age (in the time domain) or area (in a spatial domain), so long as the underlying mechanistic relationship holds. Indeed, this relationship is fundamental to the so-called species-area effect (MacArthur and Wilson 1967). "Unnatural" habitat dynamics--such as those mediated by forest management, anthropogenic disturbance, or unusual site conditions--detract from the predictive relationship by degrading the correspondence between stand age (or area) and microhabitat diversity. This is borne out by empirical studies that have shown that measures of habitat diversity can explain variation in avian diversity not explained by forest area (i.e., as significant partial correlations between habitat diversity and bird species diversity: Freemark and Merriam 1986, Lynch and Whigham 1984). The counterpoint to this is the caveat that such predictive relationships cannot be extrapolated to forest stands for which the habitat-area (age) correspondence does not hold. In reality, these invalid cases may be the rule for most habitats in developed landscapes; there are discouragingly few "natural" habitats remaining.

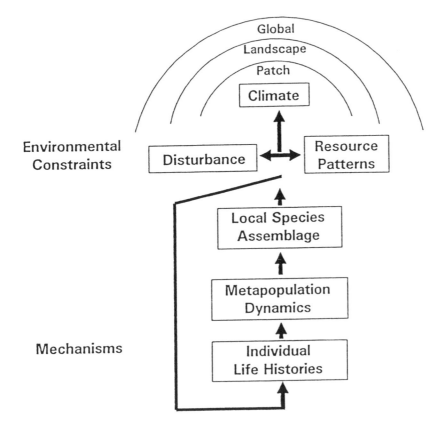

Figure 1. Avian metapopulation dynamics as a consequence of higher-level constraints acting on the demographics of individual birds.

Implications. The scaling lesson here is straightforward. At the scale of the forest stand (habitat patch) there is a mechanistic relationship that can lead to valid and accurate predictions of stand-level biodiversity as a function of stand age or area. But extrapolating such predictions under conditions of environmental change (managed forests, or a change in the relation between habitat structure and resources needed for breeding success) obviates the correlative basis for prediction. Simple extrapolations would be invalid and misleading.

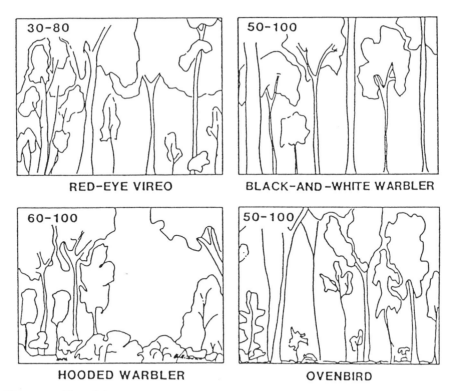

Figure 2. The niche gestalt of birds, as the characteristic vegetation profiles associated with breeding territories (examples redrawn from James 1971).

2. Landscape Pattern and Avian Diversity

As important as microhabitat patterns may be in explaining bird community structure, it has become evident over the past two decades that higher-level phenomena also constrain bird communities. The spatial and temporal patterning of landscapes accounts for additional variation in bird community characteristics. Spatial landscape metrics found to be strongly associated with bird species demography include: size distribution of suitable habitat patches (Forman *et al.* 1976, Ambuel and Temple 1983, Freemark and Merriam 1986, Robbins *et al.* 1989, Laurance 1990); habitat isolation (MacClintock *et al.* 1977, Lynch and Whitcomb 1978, Urban *et al.* 1988); boundary characteristics (Kroodsma 1982, Gates and Gysel 1978, Brittingham and Temple 1983, Wilcove 1985, Harris 1988); patch juxtapositioning (Harris 1984); and habitat diversity (Roth 1976). The effects of change in these metrics over time has been less studied but is likely to be of considerable importance (Franklin and Forman 1987).

The links between bird communities and landscape patterns are best known for forested lands undergoing fragmentation by non forest land uses. Forest fragmentation reduces total forest area, decreases mean patch size, increases the proportion of forest edge, and increases patch isolation (Sharpe *et al.* 1981, Franklin and Forman 1987). Whitcomb *et al.* (1981) identified specific life-history traits that rendered bird species especially vulnerable to forest fragmentation (Figure 4). Many of these species are

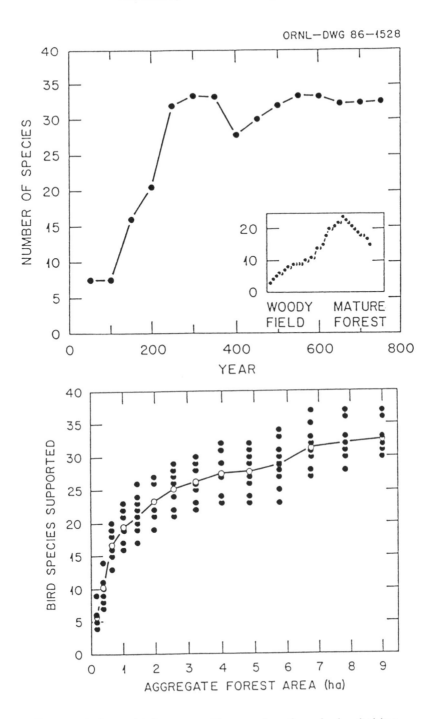

Figure 3. Patterns in forest bird communities as a function of microhabitat dynamics: (a) successional trends in species richness, (b) a species-area effect (from Urban and Smith 1989).

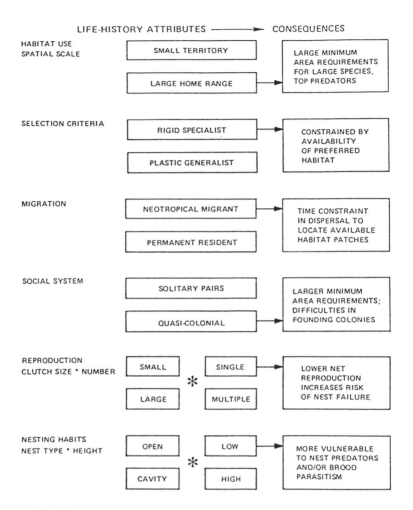

Figure 4. Life-history traits conferring sensitivity to forest fragmentation, for bird species of the eastern United States (redrawn from Whitcom et al. 1981).

suffering regional population declines across fragmented portions of the EDF (Terborgh 1989).

This fragmentation model has been documented in some other biomes around the world (Lovejoy *et al.* 1986, Laurance 1991) and is being applied increasingly to the management of biomes for which local data are not available (Harris 1984). But forest fragmentation is only one possible trajectory of landscape change (Urban *et al.* 1988, Hansen *et al.* 1991, Hansen and Urban in press). Landscape patterns and the suites of life histories represented in animal communities vary among geographical locations. Hence, animal community response to land use or climate change is apt to vary geographically.

In this vignette we summarize studies that compare avian life histories and species sensitivity to landscape change for two biomes. We also contrast bird community dynamics under forest fragmentation with those under two other common trajectories of landscape change.

Avian Life Histories in Different Biomes

The life-history traits of a species are a product of natural selection and other evolutionary forces and thus are shaped by long-term environmental, demographic, and genetic factors (Lande 1982). Because these factors differ between geographic locations, the suite of life histories represented in local animal communities are likely to differ between places. Consequently, communities from distinct biomes should differ in response to a given trajectory of landscape change in ways predictable based on the life-history traits of each community.

Table 1. Number of species and percent of total species (in parentheses) represented in various life-history guilds for Pacific Northwest (PNW) and Eastern Deciduous forest (EDF) avifaunas. The guilds are not necessarily mutually exclusive. From Hansen and Urban (in press).

Guild	Number of Bird Species:	
	PNW	EDF
Edge specialist or small forest patch associate	4 (5.3%)	27 (36.0%)
Patch interior or large forest patch associate	14 (18.7%)	27 (36.0%)
Large tree, snag or fallen tree associate	24 (32.0%)	20 (26.7%)
Neotropical migrant, closed-forest specialist, open nests near ground (0-3 m) and low reproductive effort (< = 6 eggs/yr)	0 (0.0%)	8 (10.7%)
Carnivorous, closed-forest specialist and large territory size (40 ha)	5 (6.7%)	3 (4.0%)

Hansen and Urban (in press) performed an initial test of this hypothesis by comparing the representation of various life-history guilds in avifaunas from the Eastern Deciduous forest (EDF) and Pacific Northwestern (PNW) conifer forests of North America. The EDF avifauna included many more species associated either with forest edges and small habitat patches or with habitat interiors and large habitat patches than did the PNW community; conversely, predators requiring large forest tracts were better represented in the PNW (Table 1). The life-history guild identified by Whitcomb *et al.* (1981) as being highly sensitive to forest fragmentation included eight species in the EDF but none in the PNW.

The relative sensitivity of each community to changing habitat patterns was evaluated by assigning each species a sensitivity index based on its life-history traits. The EDF avifauna in total was found to have a higher index of sensitivity to both forest fragmentation and to landscape change in general (Figure 5).

The validity of using life histories to predict community response to landscape change was supported by the fact that the sensitivity scores for PNW species correlated significantly with independent data on species population trends. As expected, the species most sensitive to landscape change underwent population increases or decreases during this period of dramatic landscape change in the PNW.

This analysis provided evidence that life-history traits are likely to differ among communities and that these differences can cause communities from distinct geographic locations to respond uniquely to a given landscape trajectory. An obvious implication is that conservation strategies should be uniquely tailored to a region based on the types of life-history attributes represented in the community. The challenge in parts of the EDF, for example, is to increase the abundance of forest interior habitats to benefit the large groups of forest-dwelling neotropical migrants. The focus in the PNW should be on maintaining: natural microhabitats, large tracts of forest for forest predators, and sufficiently large openings for open-canopy specialists.

Avian Diversity Under Differing Landscape Trajectories

Because landscape patterns and the environmental constraints imposed by them vary among geographic locations, attention to local landscape patterns is essential to predicting bird community dynamics. Hansen *et al.* (1991) illustrated this point by simulating bird habitat diversity under three common paths of landscape change in human-dominated systems: forest fragmentation, conversion of natural forest to managed plantation, and abandonment of agricultural land (termed *deprise agricole* by the French) (Figure 6). Under forest fragmentation, total forest area was reduced over the course of the simulation; the forest conversion scenario maintained total forest area but converted it to plantations. Under *deprise agricole,* agricultural lands were abandoned over an 80-yr period and then allowed to undergo natural forest dynamics, but with disturbances suppressed. Landscapes simulated under each of these trajectories were classified in term of suitability as habitat for bird species in the PNW avifauna based on four life-history traits (seral stage association, microhabitat requirements, response to forest edges, and territory size). In all cases the model assumed that microhabitat features typical of natural old-growth forest (large trees, snags, fallen trees) were present only in natural stands more than 100 years old, and that these microhabitat features were absent in non forest lands and plantations.

Interestingly, the exercise revealed that even subtle differences in the spatial scaling of landscapes can have strong effects on habitat suitability. Landscape patterns under the forest fragmentation and forest conversion scenarios were similar in that some of the seral stages found in the initial "natural" landscape (natural open-canopy stands and old growth) were lost after only a few decades. The scenarios differed in that total and interior forest area decreased under fragmentation until no forest remained, while forest area remained as high at the end of the conversion to managed forest as in the initial natural landscape.

Under *deprise agricole,* agricultural lands were allowed to gradually undergo forest development over an 80-year period. Thereafter, disturbance was omitted and forest succession resulted in a mixture of mature and old-growth stands after 220 years.

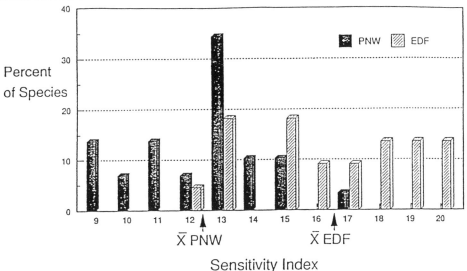

Figure 5. Frequency distributions of scores for sensitivity to forest fragmentation, based on life history-traits, of forest-dwelling bird species from the Pacific Northwest (PNW) and Eastern Deciduous Forest (EDF) avifaunas (from Hansen and Urban, in press).

Some suitable habitat existed in the initial natural landscape for all but 1 of the 51 bird species modeled (Figure 7). Habitats for several species were lost under both forest fragmentation and forest conversion, largely due to the loss of both late seral stages and natural microhabitats. There were important differences, however, in habitat richness between the forest fragmentation and forest conversion scenarios. The abundance of forest cover in the latter maintained habitats for nine more species than did the former. Among these species were some requiring intermediate sized patches of forest interior.

Under *deprise agricole,* habitat richness gradually increased as natural microhabitats and late-successional habitats became available. But habitat diversity in this run did not achieve that of the natural landscape because, under the suppression of natural disturbance, open-canopy and edge habitats were not as abundant as in the natural landscape.

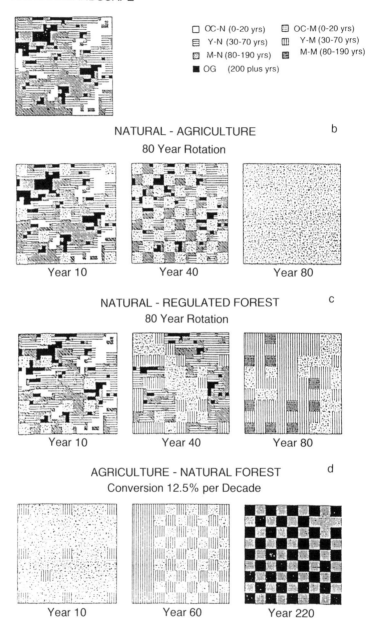

Figure 6. Maps of the simulated landscapes at various time steps. The natural landscape (a) represents year 220 under the fire regime described for a watershed in the Central Oregon Cascade Mountains. The forest fragmentation (b) and conversion to regulated forest (c) runs were initialized with map (a) and subjected to a checkerboard timber harvest regime with an 80-year rotation. The simulation of deprise agricole (d) started as non forest land, underwent forest regeneration for 80 years, forest development continued for 140 years more. From Hansen et al. (1991).

This modeling effort, simple as it was, demonstrated the important role that the type and geometry of habitats across landscapes can play in determining animal community characteristics.

Implications. These studies provide support for the paradigm that local landscape patterns fuel differential responses in animal species according to their individual

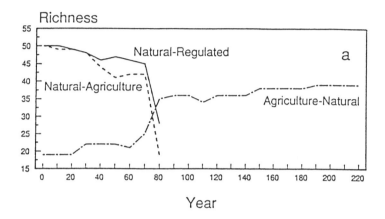

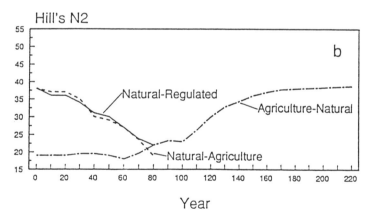

Figure 7. Richness (a) and diversity (b) of habitats for Pacific Northwest bird species over the three simulated landscape trajectories. From Hansen et al. (1991).

life-history strategies, and thus set constraints on the characteristics of animal communities. Both the life-history strategies represented in an avifauna and landscape patterns differ with geographic location. Consequently, it would appear invalid to extrapolate knowledge of animal/landscape relationships to regions or environmental regimes outside the domain of the calibration data. Rather, information on both local landscape context and local explanatory mechanisms are necessary to predict the consequences of land use or climate change.

3. Continental-Scale Patterns and Species Distribution

In this vignette we illustrate how an understanding of the large-scale movement patterns and biogeography of a single species must consider life-history traits and local site phenomena (habitat availability) as well as larger-scale constraints imposed by subcontinental air mass dynamics.

Quelea Life History and Local Patterns. The red-billed quelea (*Quelea quelea*) is a colonial, granivorous weaver-bird (Ploceinae: Ploceidae) found throughout the savannas of sub-Saharan Africa (Ward 1973). Quelea are quite gregarious throughout

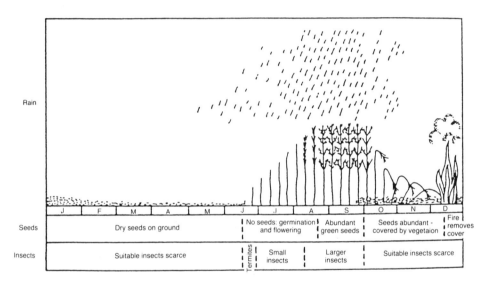

Figure 8. Idealized food availability for Quelea quelea (from Ward 1965a,b,1971).

the year and their dense breeding colonies may often include several hundred thousand pairs and cover a few tens of hectares; colonies in excess of one million pairs are not uncommon (Morel *et al.* 1957, Ward 1965a,b). As a result of its perceived status as an agricultural pest (Bruggers and Elliott 1989, Ward 1979), the red-billed quelea is one of the best studied birds in Africa (Elliott 1979, 1989; Katz 1974, 1976; Ward 1965a; Wiens and Dyer 1977).

Quelea feed heavily on grass seeds and also consume insects, particularly during the breeding season. Their diet is generally dominated by a few species of annual grasses (Ward 1965b), which tend to occur in dense stands on clay soils in swales, alluvial fans, and dry lake beds (Ward 1965a, Westoby 1980, Walker 1981). After extended drought and/or under intense grazing pressure, annuals may also dominate other sites for a time (Westoby 1980, Walker 1981). The existence of these stands of annual grasses are thus dependent on a unique disturbance regime which is driven by a combination of factors including rainfall regime, soils, geomorphology and grazing.

The phenology of annual grasses--hence quelea habitat--is driven by seasonal rains, and quelea has a life-history strategy that is geared to these dynamics (Figure 8).

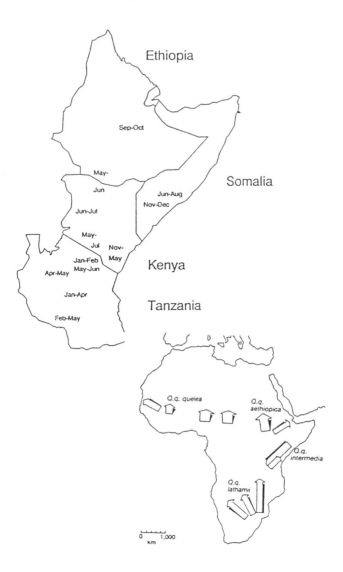

Figure 9. The spatial and temporal distribution of active breeding colonies of Quelea quelea in East Africa, (a) as positions of colonies at dates of known breeding (redrawn from Elliott 1990), and (b) as migration patterns (from Ward 1971).

The entire breeding cycle, from colony initiation to independence of the young, is completed in only six weeks (Ward 1965b; Jones and Ward 1876, 1979). Quelea are highly nomadic (Figure 9). An individual may nest two or possibly three times during a single season at locations which may be tens to hundreds of kilometers apart (Ward 1971, Jaeger *et al.* 1986, Thompson 1988, Elliott 1990, Jones 1990). The nomadic movement patterns, opportunistic nesting and rapid nesting cycle are adaptations which permit the red-billed quelea to exploit this shifting mosaic.

Continental-scale Constraints. On the basis of local observations from a few fixed points on the landscape, the arrival and departure of huge flocks of quelea appear random. A broader overview demonstrates that these movements are driven by large-scale atmospheric circulation patterns. In Africa, the seasonal rains are associated with the passage of an atmospheric circulation feature referred to as the Intertropical Convergence Zone (ITCZ). The ITCZ represents the boundary between northern- and southern-hemisphere air masses and is thus viewed as the meteorological equator (Figure 10).

The ITCZ is a very weak air mass convergence zone. The rainfall produced within the ITCZ is convective in origin, occurs in relatively small $(1\text{-}2 \text{ km}^2)$ patches, and its distribution is highly unpredictable both within and between years. The convective rainfall cells embedded within the ITCZ and the seasonal movement of the ITCZ over the African continent produce the shifting mosaic of resource patches which force the red-billed quelea to move across the landscape.

A dynamic regional overview of breeding habitat availability for the red-billed quelea can be obtained by directly monitoring vegetation dynamics using coarse-resolution satellite data (Wallin 1990, Wallin *et al.* 1991). Data provided by the Advanced Very High Resolution Radiometer (AVHRR) have a suitable spatial resolution (4 km, which approximates the minimal size of the foraging zone around a breeding colony, Bruggers *et al.* 1983) and a temporal resolution (daily) which facilitates monitoring of the rapid green-up and senescence of the savanna. Data provided by the AVHRR sensor can be used to calculate the Normalized Difference Vegetation Index (NDVI), which has been shown to be well correlated with a number of important plant processes (Sellers 1985, 1987; Tucker and Sellers 1986).

Wallin used discriminant function analysis of NDVI data to develop a classification function to predict quelea habitat suitability at two-week intervals over the subsaharan African subcontinent (Wallin 1990, Wallin *et al.* 1991). Based on 200 colonies in Tanzania, the classification was 76% accurate, and available data suggest that this approach captures the regional pattern of habitat availability in space and time. We should emphasize that this data-intensive approach predicts only where colonies might be; there is still a stochastic element as to exactly which available sites will be occupied at any given time.

Implications. Predicting quelea response to global change would involve the consideration species biology as well as environmental factors at several scales. At the local scale, an increase in the reliability of rainfall may result in an increase in the amount of perennial grasses and a decrease in the annual grasses upon which quelea depend. More reliable rainfall could also result in an increase in the populations of resident predators or granivorous competitors. A decrease in the duration of the rains could decrease the window of opportunity for nesting to the extent that quelea would no longer able to complete their nesting cycle. Finally, changes in grazing pressure keyed on altered grassland dynamics could also have an impact on quelea habitat. While it is feasible to attend these relationships in sufficient detail to make predictions at the subcontinental scale, such an approach is logistically daunting and very data-intensive.

4. Biosphere Reserves and Global Change

Our first three vignettes suggest that species life-history traits interact with environmental constraints at multiple scales (local, landscape, and continental) to generate patterns in species abundance and diversity. How, then, do we reconcile the richness of these relationships to a concern for global biodiversity? Clearly, the approach we have illustrated thus far would be logistically overwhelming if extended to many species and large scales. An alternative is the coarse-filter approach (Hunter *et al.* 1988), wherein representative samples of ecosystems are maintained in the hope that their internal diversity will be preserved as well. This is the basis for biosphere reserve

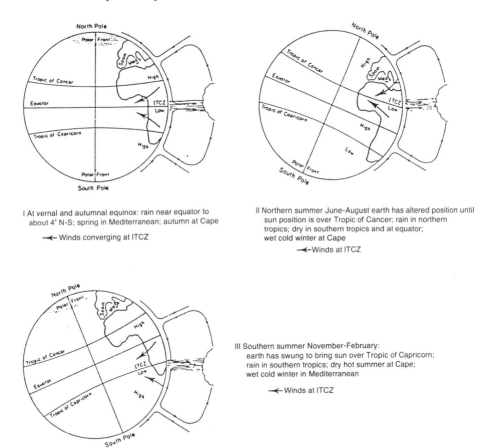

I At vernal and autumnal equinox: rain near equator to
about 4° N-S; spring in Mediterranean; autumn at Cape

◄— Winds converging at ITCZ

II Northern summer June-August earth has altered position until
sun position is over Tropic of Cancer; rain in northern
tropics; dry in southern tropics and at equator;
wet cold winter at Cape

◄—Winds at ITCZ

III Southern summer November-February:
earth has swung to bring sun over Tropic of Capricorn;
rain in southern tropics; dry hot summer at Cape;
wet cold winter in Mediterranean

◄— Winds at ITCZ

Figure 10. Movement of the Intertropical Convergence Zone (from Flohn 1987).

systems. In this last vignette we consider the implications of global climate change to a static network of biosphere reserves.

Climate-change scenarios suggest that the geographic extent of vegetation types might shift through the stationary boundaries of established nature reserves. This translocation of vegetation ranges could act to further fragment vestiges of protected species habitat. Local extinctions of reserve species occur through either direct physiological reactions of species to changed climatic conditions or through changes in interspecies competition due to alterations in the composition of species across the landscape (Figure 11) (Peters and Darling 1985, Noss 1987, Hunter et al. 1988).

Predicting the potential effects of climatic shifts on reserve systems requires a geographically referenced vegetation model which can be manipulated directly to represent the potential impacts expected from different climate-change scenarios. Coarse-resolution associations between vegetation zones or biotic provinces and climatic patterns are well established in ecology. Assuming that extant vegetation is more-or-less at equilibrium with current climate, correlative models of climate-vegetation associations can be developed (Holdridge 1967, Box 1981). From these types of models, potential shifts in vegetation pattern can be projected from general circulation model (GCM) experiments conducted by climatological research groups (Emanuel *et al.* 1985, Shugart 1990, Smith *et al.* 1990).

One such approach has involved the development of a global climate data base which can be used to calculate the spatial distribution of major ecoclimatic types (biotic provinces or biomes) and then track potential changes in the distribution of these types due to various climate-change scenarios. The model includes current climate parameters for a $0.5° \times 0.5°$ latitude and longitude grid (Emanuel *et al.* 1985, Leemans and Cramer 1990, Smith *et al.* 1990). The climatic data from this grid were then used to predict ecoclimatic regions using the Holdridge life-zone approach (Holdridge 1967). The Holdridge classification system is a correlative ecoclimatic model which relates general vegetation associations to indices of biotemperature and precipitation. The simplicity of the Holdridge method makes this approach very straightforward to use in conjunction with the output from global circulation models, for projecting large-scale shifts in vegetation patterns expected from climatic change.

Climate-change scenarios derived from four general circulation models were used to alter the baseline climatic grid to predict the expected equilibrium vegetation conditions projected under climate change. All four scenarios produced significant shifts in major biome types on a global scale (Table 2) (Smith *et al.* 1990, Monserud 1990). A large proportion of these changes were in northern latitudes, reflecting the greater temperature increases predicted in these regions.

As a first assessment of the potential impacts of climate change on a global system of nature reserves, a distribution of Man and the Biosphere (MAB) reserve sites was geo-referenced into the global grid vegetation model. Halpin and Smith, in prep.; Smith *et al.* 1990). Each reserve was assigned a present Holdridge life zone classification, with changes in life zone occurring as a result of the climatic changes predicted from the four GCM scenarios recorded as potential impacts.

The four scenarios could have a significant impact on global reserve systems (Figure 12). From a global perspective, a comparison of the percent of reserves impacted to the total area of terrestrial life zone change for each GCM scenario reveals a significant bias in impacts of MAB reserves relative to total global vegetation change. The numbers of reserves impacted by climatic change is from 12% (OSU) to 30% (UKMO) higher than the amount of total vegetation change calculated for each scenario. This bias is most likely attributable to the non uniform distribution of global reserves. A disproportionate number of reserves in the MAB system are located in the northern latitudes, specifically in Europe and North America. This geographic bias of established reserve systems to the more developed countries of the northern hemisphere puts a larger numerical distribution of reserves in regions which experience a greater magnitude of climatic change under all scenarios (Halpin and Smith, in prep.).

In general, global-scale analysis of the potential impacts of anthropogenic climate change on nature reserves offers insights into the general distribution of impacts and identifies those types suffering the highest potential loss of protection (Halpin and Smith, in prep.). On the other hand, global-scale analysis fails to offer detailed information distinguishing potential impacts within reserve areas and smaller management regions. The analysis also sacrifices the details of relationships more subtle than the simple correlative climate-vegetation model.

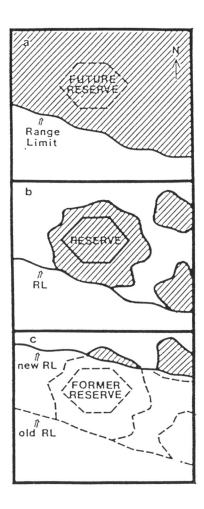

Figure 11. Potential effects of shifting climatic ranges on species occupation of a hypothetical nature reserve: (a) original species distribution; (b) species distribution after human fragmentation of the landscape; (c) species distribution after climatic shift (from Peters and Darling 1985).

Table 2. Percentage of terrestrial land area showing a shift in life zone under   changed climate (from Smith *et al.* 1990).

Scenario[1]	Percent of Land Area Changed
GFDL	48.0
GISS	44.3
OSU	39.4
UKMO	55.0

[1] Scenarios are 2x CO_2 climate-change scenarios provided by the following general circulation models: Geophysical Fluid Dynamics Lab (GFDL), Goddard Institute of Space Studies (GISS), Oregon State University (OSU), and United Kingdom Met Office (UKMO); all as provided by the U.S. Environmental Protection Agency (with no liability or endorsement implied).

Implications. Interpreting the potential impacts of climate change for nature reserves poses some unique problems. Some reserves are selected because they represent archetypal examples of the general ecological region in which they are located; on the other hand, some are selected for the opposite reason: because they represent atypical associations of species or relics of past climates. These alternative criteria in reserve selection become problematic when using coarse-resolution global-scale models, which average climate and vegetation characteristics over large spatial areas. Such models fail to distinguish the current characteristics of unusual habitats (Halpin and Smith, in prep.), and thus cannot be used confidently to project vegetation response to environmental change.

Conclusions

In closing, we should emphasize the power of life-history mechanisms and constraining context as explanatory principles at whatever scale. So long as these hold up, simple correlative relationships can be derived to provide valid predictions about biodiversity patterns in space or time. If either the mechanisms or the constraints are changed, correlative predictions may no longer be valid.

Extrapolating from one system to another can lead to confusion or misleading predictions if the systems differ in their mechanisms or context. This can be a real danger in global-change research, since we are constantly on the edge of our knowledge base and limited by available data.

At increasingly larger scales, and especially at the global scale, we come to a dichotomy in possible approaches to managing biodiversity. A mechanistic approach based on life-history traits and environmental constraints can provide a powerful predictive model, which may be applied from local to regional scales. Beyond this scale, the details of this approach become logistically daunting. This approach might still be recommended for certain well-studied focal species (e.g., endangered species) but it

Number of Sites Impacted

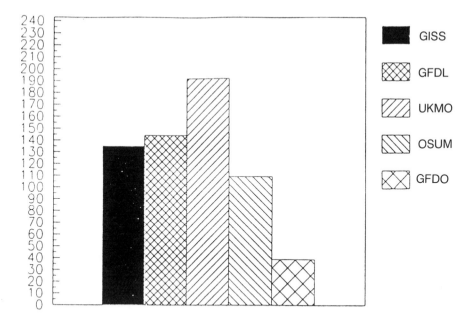

Figure 12. Percent of Man and the Biosphere Reserves Impacted under climate change scenarios.

cannot be recommended as a general solution. Coarse-scale approaches based on life forms or type vegetation assemblages and correlative models can be applied at larger scales, at some loss of information and confidence. A challenge to researchers in conservation biology will be to develop new approaches to reconciling information at multiple scales, linking fine-scale explanatory mechanisms to global patterns in biodiversity.

References

Allen, T.F.H. and T.B. Starr. *Hierarchy: perspectives for ecological complexity.* Chicago: Univ. Chicago Press.

Ambuel, B. and S.A. Temple. 1983. Area-dependent changes in bird communities and vegetation of southern Wisconsin woodlots. *Ecology* 64: 1057-1068.

Box, E.O. 1981. *Macroclimate and plant forms: an introduction in predictive modeling of phytogeography.* The Hague: Junk.

Brittingham, M. and S.A. Temple. 1983. Have cowbirds caused forest songbirds to decline? *BioScience* 33: 31-35.

Bruggers, R.L. and C.C.H. Elliott (eds.). 1989. *Quelea quelea: Africa's bird pest.* New York: Oxford Univ. Press.

Bruggers, R.L., M.M. Jaeger and J.B. Bourassa, 1983. The application of radiotelemetry for locating and controlling concentrations of Red-billed Quelea in Africa. *Tropical Pest Management* 29: 27-32.

Elliott, C.C.H. 1979. The harvest time method as a means of avoiding quelea damage to irrigated rice in Chad/Cameroun. *J. Appl. Ecol.* 16: 23-35.

Elliott, C.C.H. 1989. The pest status of the quelea. In R.L. Bruggers and C.C.H. Elliott (eds.), *Quelea quelea: Africa's Bird Pest*. pp. 17-34 . New York: Oxford Univ. Press.

Elliott, C.C.H. 1990. The migrations of the Red-billed *Quelea Quelea* quelea and their relation to crop damage. *Ibis* 132: 232-237.

Emanuel, W.R., H.H. Shugart, and M.P. Stevenson, 1985. Climatic change and the broad-scale distribution of terrestrial ecosystem complexes. *Climatic Change* 7: 29-43.

Flohn, H. 1987. Rainfall teleconnections in northern and northeastern Africa. *Theor. Appl. Climatol.* 38: 191-197.

Forman, R.T.T., A.E. Galli, and C.F. Leck, 1976. Forest size and avian diversity in New Jersey woodlots with some land use implications. *Oecologia* 26: 1-8.

Freemark, K.E., and H.G. Merriam. 1986. Importance of area and habitat heterogeneity to bird assemblages in temperate forest fragments. *Biol. Conserv.* 36: 115-141.

Franklin, J.F. and R.T.T. Forman. 1987. Creating landscape patterns by forest cutting: ecological consequences and principles. *Landscape Ecol.* 1: 5-18.

Gates, J.E. and L.W. Gysel. 1978. Avian nest dispersion and fledgling success in field-forest ecotones. *Ecology* 59: 871-883.

Graham, R.W. 1988. The role of climatic change in the design of nature reserves: the paleoecological perspective for conservation biology. *Conserv. Biol.* 24: 391-394.

Halpin, P.N. and T.M. Smith. The potential impacts of climatic change on nature reserve systems at global and regional scales. (ms in prep.)

Hansen, A.J., D.L. Urban, and B. Marks, 1991. Avian community dynamics: the interplay of landscape trajectories and species life histories. In A.J. Hansen and F. Di Castri (eds.), *Landscape boundaries: consequences for biodiversity and ecological flows*. Springer Verlag, New York. (in press).

Hansen, A.J. and D.L. Urban, 1991. Avian response to landscape pattern: the role of species life histories. *Landscape Ecol.* (in press)

Harris, L.D. 1984. *The fragmented forest*. Chicago: University of Chicago Press.

Harris, L. D. 1988. Edge effects and conservation of biotic diversity. *Conserv. Biol.* 2: 330-332.

Holdridge, L. R. 1967. *Life Zone Ecology*. Tropical Science Center, San Jose, Costa Rica.

Hunter, M.L., G.l. Jacobson Jr. and T. Webb III, 1988. Paleoecology and the coarse-filter approach to maintaining biological diversity. *Conserv. Biol.* 24: 375-385.

Jaeger, M.M., R.L. Bruggers, B.E. Johns and W.A. Erickson. 1986. Evidence of itinerant breeding of the Red-billed *Quelea Quelea* quelea in the Ethiopian Rift Valley. *Ibis* 128: 469-482.

James, F.C. 1971. Ordinations of habitat relationships among breeding birds. *Wilson Bull.* 83: 215-236.

Jones, P.J. and P. Ward. 1976. The level of reserve protein as the proximate factor controlling the timing of breeding and clutch-size in the red-billed quelea *Quelea quelea. Ibis* 118: 547-574.

Jones, P.J. and P. Ward. 1979. A physiological basis for colony desertion by red-billed queleas (*Quelea quelea*). *J. Zool., Lond.* 189: 1-19.

Jones, P.J. 1990. General aspects of Quelea migration. In R.L. Bruggers and C.C.H. Elliott (eds.), *Quelea quelea: Africa's bird pest,* pp. 102-112 . New York: Oxford Univ. Press.

Katz, P.L. 1974. A long-term approach to foraging optimization. *Amer. Nat.* 108: 758-782.

Katz, P.L. 1976. Dynamic optimization in animal feeding strategies, with application to African weaver birds. *J. Optimization Theory and Applications* 18: 395-424.

Kroodsma, R.L. 1982. Edge effect on breeding forest birds along a power-line corridor. *J. Appl. Ecol.* 19: 361-370.

Lande, R. 1982. A quantitative genetic theory of life history evolution. *Ecology* 63: 607-615.

Laurance, W.F. 1991. Ecological correlates of extinction proneness in Australian tropical rain forest mammals. *Conserv. Biol.* 5: 79-89.

Leemans, R. and W. Cramer. 1990. *The IIASA data base for mean monthly values of temperature, precipitation and cloudiness on a global terrestrial grid.* WP-90-41. Laxenburg, Austria: International Institute for Applied Systems Analysis.

Lovejoy, T.E., R.O. Bierregaard, Jr. A.B. Rylands, *et al.* 1986. Edge and other effects of isolation on Amazon forest fragments. In M.E. Soulè (ed.), *Conservation biology: the science of scarcity and diversity,* pp. 257-285. Sunderland, Mass.: Sinauer Associates.

Lynch, J.F. and D.F. Whigham. 1984. Effects of forest fragmentation on breeding bird communities in Maryland, USA. *Biol. Conserv.* 28: 287-324.

Lynch, J.F. and R.F. Whitcomb, 1978. Effects of the insularization of the eastern deciduous forest on avifaunal diversity and turnover. In A. Marmelstein (ed.), *Classification, inventory, and analysis of fish and wildlife habitat,* pp. 461-489. Washington, DC.: USDI Fish and Wildlife Service.

MacArthur, R.H. and J.W. MacArthur, 1961. On bird species diversity. *Ecology* 42: 494-498.

MacArthur, R.H. J.W. MacArthur and J. Preer, 1962. On bird species diversity. II. Prediction of bird census from habitat measurements. *Am. Nat.* 96: 167-174.

MacArthur, R.H. and E.O. Wilson. 1967. *The theory of island biogeography.* Princeton, NJ: Princeton Univ. Press.

MacClintock, R., R.F. Whitcomb, and B.L. Whitcomb, 1977. Island biogeography and "habitat islands" of eastern forest. II. Evidence for the value of corridors and minimization if isolation in preservation of biotic diversity. *American Birds* 31: 6-16.

Monserud, R.A. 1990. *Methods for Comparing Global Vegetation Maps*. WP-90-40 Austria: IIASA Laxenburg.

Morel, G., M.-Y. Morel and F. Bourliere. 1957. The blackfaced weaver bird or dioch in West Africa. *J. Bombay Nat. Hist. Soc.* 54: 811-825.

Nieuwolt, S. 1986. Agricultural drought in the tropics. *Theor. Appl. Climatol.* 37: 29-38.

Noss, R.F. 1987. Protecting natural areas in fragmented landscapes. *Natural Areas Journal* 71: 2-13.

O'Neill, R.V., D. L. DeAngelis, J. B. Waide, and T. F. H. Allen, 1986. *A hierarchical concept of the ecosystem*. Princeton, NJ: Princeton Univ. Press.

Peters, R.L. and J.D. Darling, 1985. The greenhouse effect and nature reserves. *Bioscience* 35 11: 707-717.

Robbins, C. S., D. K. Dawson and B. A. Dowell, 1989. Habitat area requirements of breeding forest birds of the middle Atlantic states. *Wildl. Monogr.* 103: 1-34.

Roth, R.R. 1976. Spatial heterogeneity and bird species diversity. *Ecology* 57: 773-782.

Schneider, S. H. 1989. Global warming: is it real and should it be a part of a global change program? In R.S. DeFries and T.F. Malone (eds.), *Global change and our common future*, pp. 209-219. Committee on Global Change, National Research Council. Washington, DC.: National Academy Press.

Sellers, P. J. 1985. Canopy reflectance, photosynthesis and transpiration. *Int. J. Remote Sensing* 6: 1335-1372.

Sellers, P. J. 1987. Canopy reflectance, photosynthesis, and transpiration II: The role of biophysics in the linearity of their interdependence. *Remote Sensing Environ.* 21: 143-183.

Smith, J.B. and D.A. Tirpak (eds.), 1989. The *potential effects of climate change on the U.S.* Washington, DC: Office of Policy, Planning, and Evaluation, U.S. Environmental Protection Agency.

Smith, T. M., H. H. Shugart, and P. N. Halpin, 1990. *Global Forests. Progress Reports on International Studies of Climate Change Impacts*. Washington, DC: U.S. Environmental Protection Agency.

Shugart, H. H. 1990. Using ecosystem models to assess potential consequences of global climatic change. *Trends in Ecology and Evolution* 59: 303-307.

Soulè, M.E. (ed.). 1986. *Conservation biology: the science of scarcity and diversity*. Sunderland, Mass.: Sinauer Associates.

Terborgh, J. 1989. *Where have all the birds gone?* Princeton, NJ: Princeton Univ. Press.

Thompson, J. J. 1988. The post-nuptial moult of *Quelea quelea* in relation to breeding in Kenya. *J. Tropical Ecol.* 4: 373-380.

Tucker, C.J. and P.J. Sellers, 1986. Satellite remote sensing of primary production. *Int. J. Remote Sensing* 7: 1395-1416.

Urban, D.L. and T.M. Smith, 1989. Microhabitat pattern and the structure of forest bird communities. *Am. Nat.* 133: 811-829.

Urban, D.L., R.V. O'Neill and H.H. Shugart, 1987. Landscape ecology. *BioScience* 37: 119-127.

Urban, D.L., H.H. Shugart, D.L. DeAngelis, and R.V. O'Neill. 1988. *Forest bird demography in a landscape mosaic.* Oak Ridge, TN: Oak Ridge National Laboratory Publication No. 2853.

Verner, J., M. L. Morrison, and C. J. Ralph (eds.). 1986. *Modeling habitat relationships of terrestrial vertebrates.* Madison: Univ. Wisconsin Press.

Walker, B. H. 1981. Is succession a viable concept in African savanna ecosystems. In D.C. West, H.H. Shugart, and D.B. Botkin (eds.), *Forest succession: concepts and applications*, pp. 431-448. New York: Springer-Verlag.

Wallin, D. O. 1990. *Habitat dynamics for an African weaver-bird: the red-billed quelea (Quelea quelea).* PhD thesis, Environmental Sciences, University of Virginia, Charlottesville, VA. 174 pp.

Wallin, D.O., C.C.H. Elliott, H.H. Shugart, C.J. Tucker and F. Wilhelmi. 1991. Satellite remote sensing of breeding habitat for an African weaver-bird. *Landscape Ecology* (in press)

Ward, P. 1965a. Feeding ecology of the black-faced dioch *Quelea quelea* in Nigeria. *Ibis* 107: 173-214.

Ward, P. 1965b. The breeding biology of the black-faced dioch *Quelea quelea* in Nigeria. *Ibis* 107: 326-349.

Ward, P. 1971. The migration patterns of *Quelea quelea* in Africa. *Ibis* 113: 275-297.

Ward, P. 1973. *Manual of techniques used in research on quelea birds.* Rome: APG:RAF/67/087 UNDP/FAO.

Ward, P. 1979. Rational strategies for the control of queleas and other migrant bird pests in Africa. *Phil. trans. R. Soc. London B.* 287:289-300.

Westoby, M. 1980. Elements of a theory of vegetation dynamics in arid rangelands. *Israel J. Botany* 28: 169-194.

Wiens, J.A., and M.I. Dyer, 1977. Assessing the potential impact of granivorous birds in ecosystems. In J. Pinowski and S.C. Kendeigh (eds.), *Granivorous birds in ecosystems*, pp. 205-266. Cambridge, U.K.: Cambridge Univ. Press.

Whitcomb, R.F., C.S. Robbins, J.F. Lynch, B.L. Whitcomb, M.K. Klimkiewicz, and D. Bystrak. 1981. Effects of forest fragmentation on avifauna of the eastern deciduous forest. In R.L. Burgess and D.M. Sharpe (eds.), *Forest island dynamics in man-dominated landscapes*, pp. 125-205. New York: Springer- Verlag.

Wilcove, D.S. 1985. Nest predation in forest tracts and the decline of migratory songbirds. *Ecology* 66:1211-1214.

17. Global change and alien invasions: implications for biodiversity and protected area management

Ian A. W. Macdonald

Introduction

There is a high probability that human-induced alterations in atmospheric composition will, over the next century, bring about changes in the Earth's environment which will be both more rapid and more extensive than at any time in the history of mankind (Bolin *et al.* 1986). Already several attempts have been made to predict the consequences of such changes for ecosystem structure and functioning (Worrest and Hader_1989) and for biodiversity (Peters 1990). The general consensus is that many organisms will be unable to adjust to the predicted rapid changes in the environment and that the rate of extinctions will increase accordingly (Ricklefs *et al.* 1990). However, the attempts to predict quantitatively the consequences of future climatic change for biodiversity often rely heavily on the record of past extinction events (Martin 1990), even though most authors readily acknowledge the unsatisfactory nature of this approach (Buffetaut 1990).

This article highlights the significance of 'alien invasions' as a factor which, although generally overlooked, is likely to massively exacerbate the impacts on biodiversity of the current rapid environmental change. Alien organisms are simply those which have been introduced to areas outside their natural distribution ranges by human activities. Alien invasion occurs when such a species spreads without intentional human assistance beyond the area of the initial introduction. These invasions are of particular significance for the maintenance of biodiversity when they occur in untransformed ecosystems (Mooney and Drake 1989).

Alien invasions: a novel challenge for the Earth's biota

The rate at which organisms are being introduced to areas outside their natural distribution ranges is now so much greater than during any previous era, that the phenomenon of alien invasion can be considered not only quantitatively but also qualitatively different from any naturally occurring biogeographic process: For example, it has been estimated that flowering plants colonized the Hawaiian archipelago unaided by Man at a rate of one species per 100 000 years. The rate of successful alien invasions has exceeded one species every two years, if this is averaged over the entire period of human occupancy (since the fourth century AD) (Loope *et al.* 1988). If, more realistically, the alien flora is deemed to have become established mainly since the arrival of Europeans in 1778 AD the rate is closer to four species per year (Loope *et al*. 1988). Comparable estimates for the Galápagos archipelago are 1/10 000 years unaided and 1/2 years since the first European arrival in 1535 AD (Loope *et al*. 1988). Alien invasions into continental ecosystems are apparently also proceeding at rates far exceeding those of natural colonizations. This has also been well documented for invasions of plants (Ashton and Mitchell 1989) and animals (Herbold and Moyle 1986) into freshwater ecosystems on the continents.

In continental terrestrial ecosystems, where natural invasions have always occurred more frequently than in insular or quasi-insular ecosystems (such as isolated freshwater lakes), alien invasions still constitute a totally novel challenge: the alien species are generally introduced from distant areas and accordingly often arrive without their co-evolved predators and parasites. By contrast, in natural invasions in such situations, the colonizing populations are usually in contact with those within the species' original distribution range. This difference is responsible for the phenomenon termed 'ecological release' which often enables an alien invader to outcompete native species with similar ecological requirements (Kruger *et al*. 1989). Although ecological release has not received much acknowledgment in theoretical analyses of the reasons for the success of alien invasions (Levin 1989), its reality is borne out daily by the widespread application of its corollary, the intentional introduction of biological control agents to reduce their competitive ability (Harris 1990).

Alien invasions are currently presenting native biotas throughout the world (Macdonald *et al*. 1989) with a challenge which has never before been present during a period of rapid climate change. Having established this point, it is now pertinent to speculate on how alien invasions might be expected to interact with global change.

Alien invasions and altered environments

In the early studies of these invasions it was often concluded that aliens could only successfully invade man-disturbed ecosystems (Elton 1958). This contention has now been shown to be false (Herbold and Moyle 1986). However, the ease with which an alien species invades an untransformed ecosystem has frequently been shown to increase where the disturbance regimes under which the native biota have evolved are modified through human agency (Macdonald *et al*. 1989). Thus, for example, historical changes in the regimes of burning and herbivory are implicated in the widespread invasion of the world's Mediterranean-climate regions and tropical oceanic islands by alien plants (Kruger *et al*. 1989; Loope and Mueller-Dombois 1989). Nutrient supply alterations and soil disturbance have been demonstrated experimentally to increase alien invasions into native Australian plant communities (Hobbs 1989). Man-induced change in the salinity regimes of western Australian salt marshes has been implicated in their successful invasion by the alien bulrush *Typha orientalis* (Zedler *et al*. 1990). Similarly, anthropogenic increases in acid precipitation have been shown to have favored the recent rapid invasion of the southern hemisphere moss *Campylopus introflexus* into calcareous dune vegetation in coastal Holland (van der Meulen *et al*. 1987) and the central Asian forb *Impatiens parviflora* into the temperate forests of Central Europe (Kornas 1990). There are many such examples, ranging from eutrophication of inland waters facilitating invasions by alien plants (Ashton and Mitchell 1989) and fishes (Arthington and Mitchell 1989) to soil disturbance by introduced pigs in Hawaii facilitating invasions by alien plants and mosquitoes (Macdonald *et al*. 1989).

The influence of global change on invasions

The global environmental changes currently underway as a consequence of the rapid increases in atmospheric concentrations of 'greenhouse' gases and of the simultaneous depletion of stratospheric ozone, are likely to cause local environmental conditions to be altered beyond the limits to which native communities have been exposed. In the past, the vast majority of alien introductions have failed to establish successful invasions: For example, of 799 alien plant species found growing in the Montpelier area of southern France at the turn of the century, at least 692 (86,6%) failed to become established (Holdgate 1986). Thus, of the 781-831 alien plant species ever recorded growing in Poland, 531 have died out and, of the remainder, only 10 (or c1% of the total number) have successfully invaded untransformed ecosystems (Kornas 1990). Elsewhere, it has been estimated that 1-2% of alien plant introductions eventually spread into untransformed areas such that they might be considered "pest species" (Holdgate 1986). Even this estimate is likely to be too high, as most failed introductions will not have been recorded (Mooney and Drake 1989).

One of the major correlates with the failure of alien invasions has been poor adaptation of the alien to local climatic conditions. It has now become one of the few general principles relating to the study of the phenomenon, that close matching of environmental conditions between the alien's original and introduced ranges is virtually a prerequisite for successful establishment (Holdgate 1986). The ecological interpretation of this relationship is unclear, but possibly lies in the generally superior competitive ability of native species under the environmental conditions under which they have evolved. In fact, given the finely-tuned relationships between organisms and their native environments that are increasingly being demonstrated, it is indeed remarkable that alien organisms, having generally evolved under very different conditions, are ever able to invade native communities. It is probable that, at least in the past, it was only the twin phenomena of ecological release and subtly altered disturbance regimes that enabled an alien organism to outcompete a native species in an apparently 'undisturbed' ecosystem. Certainly, the introduction of only one or a few of an alien species' many predators, parasites or diseases has repeatedly been shown to reduce or eliminate this competitive ability (Harris 1990).

Given a scenario of rapid change in local abiotic conditions, it is likely that native species will no longer necessarily be better adapted to the new environmental conditions than a species introduced from elsewhere. It can be predicted therefore that the proportion of alien introductions that become alien invaders will increase.

CO_2 fertilization and photosynthesis in invasive plants

The one component of the current global change which is currently not disputed is the rapid increase in carbon dioxide concentrations in the atmosphere (Bolin *et al*. 1986). It is generally accepted that mean global concentrations have increased from some 280 ppmv in the 1700s to about 349 ppmv by the end of the 1980s (Fajer 1989). This constitutes a 25% increase above the highest levels recorded in the 18th century. Regardless of the associated impacts on climate, this CO_2 increase has predictable effects on the photosynthetic capacity of plants (Pearcy and Ehleringer 1984). One of these predictions is that plant species utilizing the "C_3" and "CAM" photosynthetic pathways should increase their photosynthetic efficiencies, at least where the plants are not stressed for water, whereas plants utilizing the "C_4" pathway should not (Fajer 1989).

It is of interest to see whether the hypothesis of alien species being advantaged by global environmental changes in the future, is borne out by an examination of which photosynthetic pathways characterize those alien plant species which have succeeded to date. Unfortunately, global inventories of alien invaders or of photosynthetic mechanisms in plants are lacking. Two regional analyses are available: In southern Africa (Whiting *et al*. 1986) it was found that of 140 "aggressive alien invaders" for which the mechanism of photosynthesis was known, 128 were C_3, 6 were CAM and 6 were C_4. When only the 23 "major" invaders were considered, 20 were C_3, 3 were CAM and none was a C_4 species . Although no comprehensive survey of photosynthetic mechanisms in the region's native flora has been carried out, it is likely that the percentage of native species utilizing the C_4 pathway is higher than the 4.3% they constitute of the aggressive alien flora and it is certain this percentage is higher than the 0% they constitute of the major invaders. If the distribution of invasions within the region is taken into account, then the proportion of C_4 species in the alien flora is very much lower than it is in the native flora of areas where the level of invasion is currently highest. Vogel (1978; Vogel *et al*. 1978) showed that over most of South Africa more than 90% of grass species show the C_4 metabolism. Grasses alone constitute 4% of the South African flora (Wells *et al*. 1983) and there are numerous other families present in the flora which have members exhibiting the C_4 pathway. Werger and Ellis (1981), in the only South African study which has attempted to characterize the entire flora of specific areas according to photosynthetic pathways, show that in the eastern mesic zone at about $30°30'S$ the contribution of C_4 species to the flora is 15-30% and to the vegetation about 40-60%. It is these mesic areas on the eastern seaboard that have some of the subcontinent's most diverse and successful invasive alien floras (Macdonald 1985).

In Australia, an analysis of the "noxious plants" of the state of Victoria showed that only five of the 83 species for which photosynthetic mechanisms were known are C_4 species (Newsome and Noble 1986). Although the comparable statistics for the state's native flora are not presented, the point is made that this proportion is much lower than would have been predicted using the conventional notion that C_4 species are competitively superior.

Similarly, the invasive alien flora of early secondary successions in tropical India has been found to be dominated by C_3 species whereas the native flora in such sites is primarily made up of C_4 species (Ramakrishnan and Vitousek 1989). This apparent dominance of invasive alien floras by C_3 species is in marked contrast to the situation in the weed floras of arable lands: In such situations where, presumably, interspecific competition is less intense, lists of weeds are dominated by C_4 species. Thus, for example, in the "world's worst weeds" 14 of the 18 top ranked weeds were C_4 species (Newsome and Noble 1986). The relative paucity of C_4 species in invasive alien floras might be a consequence of these having been selected in closed communities where interplant competition can be presumed to be more intense. Under such circumstances, the influence of recent CO_2 fertilization might have shifted the competitive advantage in favor of C_3 species.

Many of the 'weed' species of cultivated lands were derived from the available species pool long before anthropogenic CO_2 fertilization of the global atmosphere could possibly have been a significant selective factor, most are "eu-apophyta" and "archaeophyta" *sensu* Kornas (1990).

Thus, what little is known about the distribution of photosynthetic mechanisms in the more recent invasive alien floras is consistent with the hypothesis that global change is favoring the successful species. However, it must be emphasized that this is simply an examination of a concept and that more detailed analyses are required to eliminate the possibility that it is factors correlated with these photosynthetic pathways that are being selected for, rather than the photosynthetic metabolism itself.

Implications for biodiversity

If the overall contention, that global change is going to promote alien invasions, is accepted, what are the implications for biodiversity? First, it should be noted that such invasions have already been responsible for a large proportion of all the extinctions that are known to have occurred during the historical period (Macdonald *et al.* 1989). Secondly, there is the increasing localization of much of the world's native biota in small quasi-insular protected areas, often having long boundaries with transformed ecosystems in which alien species are numerous. Thirdly, there is no hard evidence that the rate of introduction of potentially invasive alien species is decreasing anywhere in the world, in fact the converse might well be true, at least for fishes, plants and insects. In one apparent exception to this generalization, Kornas states that for Central Europe "it seems that there has already been a marked decrease in new arrivals [of alien plant species] in the recent years" and he predicts that this trend will continue in the future, possibly as a result of most potentially invasive taxa already having been introduced to this area (Kornas 1990). He does not, however, provide statistical data in support of this contention and it is possible that many recent invasions have not yet been detected, particularly as most invasive species remain in very low numbers for some time after their introduction to a new area.

Fourthly, it is now widely recognized that a successful alien invasion often facilitates further invasions, i.e. the process is self-reinforcing (Cheke 1989). Finally, it should be noted that it is not a random subset of the world's biota that constitutes the invasive alien species: Instead, it is a relatively small number of species which are coming to dominate ecosystems in totally different biogeographic regions. The trend is one of a net loss in global biodiversity; a homogenization of the Earth's biota, as was recognized early on by Charles Elton (1958).

Implications for plant extinctions

In paleo-ecological analyses of past extinction episodes, most of which have been attributed to rapid environmental change, one of the conclusions has been that plants have shown much reduced extinction rates relative to animals (Traverse 1990). This is likely to change radically in the presence of alien species: For example, it is estimated that, already, some 900 plant species in the Cape Floral Kingdom are threatened with extinction, at least in part as result of alien tree invasions (Macdonald *et al*. 1989). In this same area, approximately 1000 myrmecochorous plant species could be threatened if the already established alien ant *Iridomyrmex humilis*, which does not bury plant seeds, continues to expand into native plant communities where it replaces native seed-dispersing ants. Thus, somewhere between 12% and 25% of the 7 316 plant species known from the "core area" of the fynbos biome, which includes the Cape Floristic Kingdom (Cowling *et al*. 1989), are estimated to be potentially at risk of extinction as a result of just these two types of alien invasion. There are numerous other alien taxa invading the biome, all of which add to the number of endemic plant species threatened with extinction; invasions are now a major factor in this area's very poor conservation outlook.

This situation is not unique to the Cape region; similarly pervasive plant and animal invasions are known to have displaced native plant species from most of their original ranges in other Mediterranean-type regions, temperate grasslands and islands. Although most of the affected species still maintain small remnant populations, it would only require an increase in the intensity of the invasion, and/or the relaxation of current control measures for many to become extinct. Numerous local populations of such species have already become extinct in the Cape as a direct consequence of alien plant invasions.

Historically, some of the most devastating alien invasions of continental biotas have been by micro-organisms (Macdonald *et al*. 1989). These invasions have often come close to completely eliminating susceptible plant and animal species, and have frequently brought about local extinctions of either the affected species itself or of other species closely dependant upon such a species. The distributions of many such microorganisms are closely controlled by climate. One of the unfortunate implications of a rapid climate change is that certain areas formerly unsuitable for an invasive microorganism will suddenly become prone to invasion. These areas could in certain instances be providing the last refugia for highly susceptible species. The plant pathogenic fungus, *Phytophthora cinnamomi*, which currently poses a major threat to native plant species in Australasia, provides a classic example of such a situation (Podger *et al*. 1990).

The plant extinction event that could be precipitated by alien invasions in the next century if control strategies are not effectively implemented, dwarfs that of the historical past and possibly those of the geological past.

Changing distributions of established alien invasions

The few predictions that have been made on the effects of global change on alien invasions have mainly concentrated on the possible changes in the ranges of established invasive alien species. This effect is likely to be substantial, particularly as global warming will make much of the what is currently the northern temperate zone invisible by more tropical species of which there are a higher diversity than there are temperate species. However, the converse effect will also manifest itself, i.e. some aliens of temperate origin will tend to show overall reductions in the ranges in which they are invasive. This is unlikely to outweigh the previous phenomenon for reasons of the relative size of temperate and tropical biotas. Notwithstanding this, I do not consider the changing ranges of established invasive taxa to be the most important likely effect of global change on the invasion process. Instead, this is likely to be the enhanced ease with which alien species, both new and established, outcompete native taxa.

Attitudes and responses to alien invasions

Is it appropriate that humankind should adopt a spectator role when faced with the prospect of a massive upsurge in alien invasions in the near future? If the historical precedent is to be followed, this might well prove to be the case: in general, the problem has not been tackled timely and effectively anywhere in the world. The few significant attempts that have been made to control invasive alien organisms for purposes of maintaining biodiversity (as distinct from those motivated primarily by agricultural and sylvicultural purposes), have generally been too little or too late.

The conservation community as a whole has, until very recently, tended to avoid addressing the problem (Ehrlich 1980). Thus, for example, in the World Conservation Strategy (IUCN 1980), although the severity of the problem is mentioned in its introductory section, the topic is not included in the formulation of international conservation priorities or checklists for action. Even perceptive conservation biologists such as Paul Ehrlich, Richard Bell, Michael Soulè and Norman Myers, although generally acknowledging that alien invasions will contribute to the current extinction episode, have not emphasized the central role they might play in the extinction process.

Very recently, information on the deleterious effects of alien invasions on island biota's has led to a sharp increase in the level of awareness of the potential significance of this phenomenon, at least amongst the world's top conservation biologists.

This welcome increase in scientific awareness has, however, been paralleled by the development of a disturbing trend in attitudes towards alien invasions: Recently, it has become fashionable to deride the informed ecologist's desire to eliminate invasive aliens wherever possible (Brideswater 1989). Instead, it has been proposed that these invasions are an inevitable corollary to modern human-induced changes in the environment and one should simply make the best of them.

The special case of protected areas

Westman (1990) has even gone so far as to suggest that, faced with the in-
evitability of rapid climatic change, conservation authorities should stop controlling
invasive alien plants in protected areas, as these might be the only species capable of
maintaining essential ecological processes in the years ahead. However, he uses as his
analogy to justify this assertion, the persistence of only a single alien grass species in
certain areas subjected to intense acid precipitation. This acidification of the environment constitutes a more radical deviation from the environmental conditions under
which the native communities evolved, than are those predicted to occur as a result of
CO_2 fertilization, global warming and stratospheric ozone depletion. It is the more
general considerations posed by these latter, globally operating, environmental perturbations which this report addresses. In environments destroyed by more localized forms
of pollution, the attitude towards alien organisms might have to be different.

It could, however, be exceedingly dangerous if conservation authorities used
impending global changes as an excuse for adopting the often expressed view that this
control is not only practically impossible but, possibly also, ecologically undesirable.
These views are generally only espoused by people who have not made a thorough
investigation of the phenomenon. By contrast, scientists who have studied the causes,
consequences and control of alien invasions are normally unanimous that such control
is not only essential, but also possible, given current technology and appropriate levels
of funding.

How can alien invasions best be contained?

The importance of detecting alien invasions early and initiating appropriate
levels of control immediately, has repeatedly been stressed (Usher 1989). The managers
of all protected areas should initiate continuous monitoring systems to detect new alien
invasions as soon as they begin. The major axes along which aliens enter a reserve must
be identified, for example, influent rivers, roadways, utility lines and areas adjacent to
or downwind of settlements.

Regarding the feasibility of control, Usher's (1989) conclusion is worth reiterat-
ing: "Scientist have developed the methods; what is needed is the political will to use
them". Soulè has further made the point that new developments in biotechnology should
enable innovative approaches to be made in the control of previously intractable alien
invaders. He predicts that, by combining existing and novel technologies, "it probably
will be possible to eliminate most exotic species in less than a decade after the initiation
of a programme" (1990). If those charged with the responsibility of conserving the
Earth's biotic diversity are to succeed, they would do well to initiate appropriate
management strategies for alien organisms now. The scientific basis for such activities
has recently been reviewed by SCOPE (the Scientific Committee on Problems of the
Environment of the International Council of Scientific Unions) (Drake *et al*. 1989).
There can be no excuse that information on the topic is insufficient for the formulation
of such strategies. Given the added dimension of a rapid change in the global environment, continuing to ignore the problem of alien invasions will almost certainly ensure

that previous predictions of the future extent of the current extinction episode are gross underestimates.

Any attempt to predict the economic consequences of alternative responses to global climatic change (Maddox 1990) must take into account either the cost of this additional loss in biodiversity as a result of enhanced alien invasions or the additional cost that will be incurred in preventing such loss. Similarly, when assessing the risks to biodiversity posed by global change it is essential that alien invasions be included in the assessment.

References

Ashton P.J. and D.S. Mitchell, 1989. Aquatic plants: Patterns and modes of invasion, attributes of invading species and assessment of control programmes. In J.A.Drake *et al.,* (eds.), *Biological Invasions: A Global Perspective* pp. 111-154. New York: Wiley.

Arthington, A. H. and D. S. Mitchell, 1986. Aquatic invading species. In R. H. Groves and J. J. Burdon (eds.), *Ecology of Biological Invasions: An Australian Perspective*, pp. 34-53. Canberra: Austrlian Academy of Sciences.

Bolin B., B.R. Doos, J. Jager and R.A. Warrick (eds.), 1986. *The Greenhouse Effect, Climatic Change and Ecosystems*. New York: Wiley.

Brideswater, F. C. 1989. Review: the ecological and management of biological invasions in southern Africa. *J. of Dendrology* 12: 47-48.

Buffetaut, E. 1990. The relevance of past mass extinctions to an understanding of current and future extinction processes. *Palaeogeog. Palaeoclim. Palaeoecol.* 82: 169.

Cheke A.S. 1987. An ecological history of the Mascarene Islands, with particular reference to extinctions and introductions of land vertebrates. In A.W. Diamond (ed.), *Studies of Mascarene Island Birds*, pp. 5-89. Cambridge: Cambridge University Press.

Cowling R.M., G.E. Gibbs-Russell, M.T. Hoffman and C. Hilton-Taylor, 1989. Patterns of plant species diversity in southern Africa. In B.J. Huntley (ed.), *Biotic Diversity in Southern Africa: Concepts and conservation*, pp. 19-50. Cape Town: Oxford University Press.

Drake J.A, H. A. Mooney, et al. (eds.), 1989. *Biological Invasions: A Global Perspective*. New York: Wiley.

Ehrlich P. 1980. The strategy of conservation, 1980-2000. In M.E. Soulè and B.A.Wilcox, (eds.) *Conservation Biology: An evolutionary-ecological perspective*, pp. 329-344. Sunderland, Mass.: Sinauer Associates.

Elton C.S. 1958. *The Ecology of invasions by Animals and Plants*. London: Methuen.

Fajer E.D. 1989. How enriched carbon dioxide environments may alter biotic systems even in the abscence of climatic changes. *Cons. Biol.* 3: 318.

Harris P. 1990. Environmental impact of introduced biological control agents. In M. Mackauer, L.E. Ehler and J. Roland (eds.) *Critical Issues in Biological Control*,pp. 289-300. Andover, Hants: Intercept.

Herbold B. and P.B. Moyle 1986. Introduced species and vacant niches. *Amer. Nat.* 128, 751-760.

Hobbs R.J. 1989. The nature and effects of disturbance relative to invasions. In J. A. Drake *et al.* (eds.), *Biological Invasions: A Global Perspective*, pp. 389-405. New York: Wiley.

Holdgate M. W., 1986. Summary and conclusions: characteristics and consequences of biological invasions. *Phil. Trans. R. Soc. Lond.* B314: 733-742.

IUCN, 1980. *World Conservation Strategy*. Gland, Switzerland: IUCN-UNEP-WWF.

Kornas J. 1990. Plant invasions in central Europe: Historical and ecological aspects. In F. Di Castri, A.J. Hansen and M. Debussche, (eds.), *Biological Invasions in Europe and the Mediterranean Basin*, pp. 19-36. Dordrecht: Kluwer Academic Publishers.

Kruger, F. J., C. J. Breytenback, I. A. W. Macdonald and D. M. Richardson, 1989. The characteristics of invaded Mediterranean-climate regions. In J. A. Drake *et al.* (eds.), *Biological Invasions: A Global Perspective*, pp. 181-213. New York: Wiley.

Levin S.A. 1989. Analysis of risk for invasions and control programs. In J.A.Drake *et al.* (eds.), *Biological Invasions: A Global Perspective*, pp. 425-435. New York: Wiley.

Loope L.L. and D. Mueller-Dombois, 1989. Characteristics of invaded islands, with special reference to Hawaii. In J.A.Drake *et al.* (eds.), *Biological Invasions: A Global Perspective*, pp. 257-280. New York: Wiley.

Loope, L.L., O. Hamann and C. P. Stone, 1988. Comparative conservation biology of oceanic archipelagos. *Bioscience* 38: 272.

Macdonald I.A.W. 1985. The Australian contribution to southern Africa's invasive alien flora: an ecological analysis. *Proc. Ecol. Soc. Australia* 14: 225-236.

Macdonald, I.A.W., L. L. Loope, M. B. Usher and O. Hamann. 1989. Wildlife conservation and the invasion of nature reserves by introduced species: a global perspective. In J.A.Drake *et al.* (eds.), *Biological Invasions: A Global Perspective*, pp. 215-255. New York: Wiley.

Maddox J. 1990. The greenhouse question (cont'd). *Nature* (Lond.) 345: 473.

Martin, P.S. 1990. 40 000 years of extinctions on the "planet of doom." *Palaeogeog. Palaeoclim. Palaeoecol.* 82, 187-201.

Mooney H.A. and J.A. Drake, 1989. Biological invasions: a SCOPE program overview. In J. A. Drake *et al.*, (eds.), *Biological Invasions: A Global Perspective*, pp.491-508. New York: Wiley.

Newsome A.E. and I.R. Noble,1986. Ecological and physiological characters of invading species. In R. H. Groves and J. J. Burdon (eds.), *Ecology of Biological Invasions: An Australian Perspective*, pp. 1-20. Canberra: Australian Academy of Sciences.

Pearcy R.W. and J. Ehleringer, 1984. Comparative ecophysiology of C3 and C4 plants. *Plant Cell Environ.* 7: 1-13.

Peters, R. L. (ed.) 1990. *Consequences of the Greenhouse effect for biological diversity*. New Haven: Yale University Press.

Podger, F. D. D. C. Mummery, C. R. Palzer and M. J. Brown, 1990. Bioclimatic analysis of the distribution of damage to native plants in Tasmania by *Phytophtora cinnamomi. Aust. J. Ecol.* 15: 281-289.

Ramakrishnan P.S. and P.M. Vitousek, 1989. Ecosystem-level processes and the consequences of biological invasions. In J.A.Drake *et al.* (eds.), *Biological Invasions: A Global Perspective*, pp. 281-300. New York: Wiley.

Ricklefs R.E. *et al.,* 1990. Biotic systems and diversity report of working group 4. Interlaken workshop for past global changes. *Palaeogeog. Palaeoclim. Palaeoecol.* 82: 159-168.

Soulè, M.E. The onslaught of allien species and other challenges in the coming decades. *Conservation Biology* 4: 233-239.

Traverse, A. 1990. Plant evolution in relation to world crises and the apparent resilience of Kingdom Plantae. *Palaeogeog. Palaeoclim. Palaeoecol.* 82: 203-211.

Usher, M.B. 1989. Ecological effects of controling invasive terrestrial vertebrates. In J.A.Drake *et al.* (eds.), *Biological Invasions: A Global Perspective*, pp. 463-489 . New York: Wiley.

van der Meulen F., H. van der Hagen and B. Kruijsen, 1987. *Campylopus introflexus*. Invasion of a moss in Dutch coastal dunes. *Proc. Kon. Ned. Akad. van Wetensch.* C 90: 73-80.

Vogel J.C. 1978. Isotopic assessment of the dietary habits of ungulates. *S. Afr. J. Sci.* 74: 298-301.

Vogel J.C., A. Fuls, and R.P. Ellis, 1978. The geographical distribution of Kranz grasses in southern Africa. *S. Afr. J. Sci.* 74, 209-215.

Wells M.J., V.M. Engelbrecht, A.A. Balsinhas and C.H. Stirton, 1983. Weed flora of southern Africa 2. Power shifts in the veld. *Bothalia* 14: 961-965.

Werger M.J.A. and R.P. Ellis, 1981. Photosynthetic pathways in the arid regions of South Africa. *Flora* 171: 64-75.

Westman W.E. 1990. Managing for biodiversity. *Bioscience* 40: 26-33.

Whiting, B. H., G. C. Bate and D. J. Erasmus, 1986. In I. A. W. Macdonald, F. J. Kruger and A. A. Ferrar (eds.), *The ecology of Biological Invasions in southern Africa*, pp. 179-188. Cape Town: Oxford University Press.

Worrest R.C. and D-P. Hader, 1989. Effect of stratospheric ozone depletion on marine organisms. *Environ. Conserv.* 16: 261-263.

Zedler J.B., E. Paling and A. McComb, 1990. Differential responses to salinity help explain the replacement of native *Juncus kraussii* by *Typha orientalis* in western Australian salt marshes. *Aust. J. Ecol.* 15: 57-72.

18. Human Aspects of biodiversity: An evolutionary perspective

Robert K. Colwell

Evolutionary inventions

The growing science of evolutionary ecology seeks an historical and functional understanding of genetic adaptations through the study of their role in suiting organisms to one another and to their abiotic environment (Colwell 1985; Price *et al*. 1991). Because they arise through natural selection, organismal adaptations represent trial-and-error solutions to ecological challenges. In this way, they resemble human inventions--minus the insight that allows us to guess, without actual trial, which modifications of previous inventions are more likely to succeed.

Civilization was founded on the adaptive "inventions" of other species, as well as our own contrivances (Colwell 1989). The domestication of food plants simply improved on the existing storage tissues of plants--seed endosperm, roots, tubers (Anderson 1967). With fiber plants, we simply improved and extracted the support tissues (linen, sisal, hemp, jute) or fibers involved in seed dispersal (cotton, kapok). The effective principles of drug plants, spices, herbs, and natural dyes rely heavily on compounds evolved by plants in protective response to the depredations of insects, mites, and diseases (Simpson and Connor-Ogorzaly 1986). The use of domesticated animals for work or transport simply exploits their existing capacity for locomotion; use of their pelts or hair or feathers for clothing mimics the role these tissues serve for their original owners; human use of animal milk exploits one of the fundamental evolutionary inventions of the mammals and our use of honey an analogous invention of bees. "Domesticated" microorganisms make possible alcoholic beverages, leavened bread, cheese, yoghurt, and soy sauce as well as industrial fermentation processes and many drugs.

True domesticates depend, to a greater or lesser degree, upon human agency for propagation or survival. The wild relatives of maize (*Zea mays*) in Mesoamerica, for example, called *teosinte* (a term that includes 3 subspecies of maize itself and three other species of *Zea*) are self-propagating (Dobley 1990). But in cultivated maize, the process of artificial selection has produced a plant with seeds (kernels) so firmly attached to the inflorescence (ear) and so tightly protected by the husks that, without the help of farmers, probably not a single living plant of domesticated maize would exist upon the earth after a few years of neglect. Like maize, many other crop species are just as dependent on us as we are on them. Animals that we regard as domesticated tend to be less dependent. Farmyard pigs (*Sus scrofa*), for example, easily revert to the behavior and even the appearance of their wild ancestors (Singer *et al*. 1984).

Yet, in traditional cultures--even in those in which domesticated species represent an essential element, wild species usually play a critical cultural and nutritional role (Denslow and Padoch 1988). The Lacandon Maya of present-day tropical Mexico, for example, clear small rain forest patches to plant mixed forest gardens of cultivars such as plantains and manioc. But in weeding these gardens, they spare seedlings of many kinds of useful wild plants that spring up from the seed bank in the soil. As the cycle of shifting agriculture proceeds, older gardens are gradually reforested with a mixture of planted woody cultivars, such as cacao, and useful native trees that have been allowed to prosper. Not only wild plants, but wild edible animals are part of the Lacandon forest garden ecosystem (Nations and Nigh 1980; Nations 1988). The evolutionary effects of cultural activities on such wild companion species of traditional cultures may sometimes be significant, but surely less so than in the transformation of wild into domesticated plants. For the most part, human use of wild species relies on the unmodified adaptive "inventions" that natural selection has produced.

Biotic and abiotic adaptations

The vast diversity of form and function among living things is largely a product of equally diverse patterns of natural selection, in time and space. At the genetic level, global diversity in organismal form and function arises from an almost unimaginable complexity of variation. The evolutionary adaptations of any species can, in principle, be divided into those that suit the organism to its abiotic environment and those that help the organism to cope with its relations with other living things. A brief discussion of the characteristics and prevalence of these two classes of adaptation in nature will lay necessary groundwork for what follows.

At the outset, it must be stressed that both abiotic and biotic selection often affect the same feature of an organism. For example, hummingbird species vary with elevation above sea level in the length of their wings relative to body mass (wingdisc-loading). At higher elevations, where the air is thin, longer wings are needed to fly efficiently clearly an example of abiotic selection. Yet, at any given elevation, species that forage for nectar among rich but widely spaced flowers (circuit foragers or "trapliners") tend have longer wings for their mass than species that defend feeding territories. The circuit foragers fly more efficiently than the territorialists, but the latter--which are better able to manoeuvre and accelerate--usually win all the fights over flowers (Feinsinger et al. 1979). These differences in relative wing length are, thus, adaptive responses to biotic selection. To add further complexity, within some hummingbird species, males and females differ in wing morphology as a result of sexual selection, arising from differential mating success (Skutch 1973).

For most plants, resources for growth are strictly abiotic--sunlight, carbon dioxide, and inorganic nutrients. (The exceptions are parasitic or carnivorous plants.) The metabolic pathways that transform these resources into living tissue inevitably produce intermediates or by-products. Yet many--perhaps most--wild plants apparently produce and retain some of these compounds in much higher than incidental concentrations, as an adaptive defense against damage or destruction by herbivores. In turn, the metabolism of herbivores, especially insects, is thus under intense selection to overcome plant defenses (Ehrlich and Raven 1964; Gilbert and Smiley 1978; Feeny 1991).

What proportion of global genetic variation--what proportion of the evolution-ary inventions of organisms--might be due to abiotic selection what proportion to biotic? Evolutionary adaptations to the non-biological environment are certainly remarkable and varied--especially in extreme environments such as deserts, hot springs, the ocean abyss, and at high latitudes. Yet, they are surely no match for the richness, diversity, and singularity of adaptations involved in biotic interactions. In addition to the generally antagonistic interactions that select for special adaptations in predators and prey, herbivores and plants, parasites and hosts, an underappreciated wealth of mutualistic associations ties together the species in natural communities, and adds immensely to the number and variety of special adaptations arising from species interactions. Were pollination mutualisms, fruit and seed dispersal by animals, nitrogen fixation symbioses, mycorrhizal associations, and digestive symbioses suddenly to disappear, biological communities everywhere would grind to a halt (Boucher *et al*. 1982). Even less ap-preciated than these relatively well-defined categories of mutualism is a large class of otherwise unclassifiable interactions that Wilson (1980) has called "non-trophic" biotic interactions.

Although I know of no quantitative data to support it, the conjecture that biotic adaptations are particularly rich in idiosyncratic detail and variety among tropical organisms would likely evoke an assenting nod from any tropical naturalist. To begin with, where such a great diversity of species coexist, the number of possible interactions rises exponentially. In addition, the low densities of many tropical organisms means that "signals" of a particular kind must be highly distinctive and particularly strong to get through the informational "noise" of competing signals. In most rain forests, for example, tree species are so diverse that the average distance between two individuals of a given species may be hundreds of meters--a space packed with a hundred or more different tree species. For efficient pollination in such an environment, flowering phenology must be coordinated and specialized pollinators must be "courted" with attractants and rewards (Baker 1978; Feinsinger 1987; Newstrom *et al*., in press). Herbivores in such a forest are faced with a complex mosaic of plant defensive chemicals. Some have responded by evolving generalized means of detoxifying their food, but many appear to have overcome the defenses of a narrow range of host plants by evolving highly special-ized metabolic adaptations and specialized sensory adaptations for finding their hosts. For insects and mites, mate-finding in the biochemical cacophony of a rain forest can also be a problem. Although some have solved it with pheromones or acoustic com-munication, others use host plants as cues for finding their own mates (Colwell 1986a,b).

Component communities in nature

In nature, the adaptations of a species to its biotic environment often focus on only a small subset of its companion species in the ecosystem. Typically, each species has strong ecological and evolutionary interactions with, at most, a few dozen others--and weak interactions with the rest of the species in its community. Such a nexus of strong interactions has been called a "component community" by Root (1973). Theoretical work (Gardner and Ashby 1970; May 1981) suggests that such ecologically-based "substruc-turing" of communities may permit a greater number of species to coexist in dynamic equilibrium, for a given level of stability in the resource supply.

The component community of an animal species would include its principal food species, of course, but also its principal enemies (predators, parasites, disease organisms), competitors, mutualists, and commensals. (Commensals, or "guests," are companion species that do neither any particular harm nor any particular good for their host.) Rain forest leaf-cutter ants (*Atta* species), for example, gather leaves sustainably from a large subset of the plants in their habitat. Since they often harvest mature leaves (Howard 1988; Wilson 1980), they probably compete strongly with relatively few other herbivores, many of which are restricted to feeding on young leaves. Leaf-cutters are subject to parasitism by a few species of specialized flies, and are host to several kinds of commensal "guest" species in their nests. Many kinds of arthropods can be found feeding in the rich organic debris of the ants' refuse heaps, but most are probably generalized scavengers.

Perhaps the greatest interspecific evolutionary effect of these ants has been on the fungus that they cultivate in the nest in underground "gardens." Although the adult ants chew up the leaf and flower material they harvest and obtain nutrients from these leaves, this activity also serves to prepare a substrate for the fungus the sole source of nutrients for leaf-cutter ant larvae is the fungus itself, and adults require enzymes that the fungus produces (Quinlan and Cherrett 1979; Howard 1988). The fungal species that leaf-cutters cultivate are highly domesticated species--so specialized that there is a one-to-one match between fungal species and *Atta* species. Just as cultivated maize can no longer reproduce without human assistance, the fungus in the ant nest cannot reproduce without the attention of the leaf-cutting ants (Stevens 1983). At most, then, a leaf cutter species interacts strongly at the ecological level with a few hundred other species (if we stick to fairly direct interactions) out of the tens of thousands or hundreds of thousands of species (no one knows for sure) in its habitat. Strong and specific evolutionary effects are likely to be even more limited.

Human component communities

Before the advent of industrialized society, the human "component community" was likewise restricted to a relatively small subset of the species in our habitats, if contemporary hunter-gatherer, pastoralist. and traditional agricultural societies are any indication. While hunter-gatherer societies are remarkable for the wide variety of wild species they exploit, the number on which their activities exert strong ecological or evolutionary effects is certainly much smaller (Hames and Vickers 1983).

In traditional agricultural societies, the true domesticates--crop plants and dooryard animals--are deeply entwined with human culture, both ecologically and in their evolutionary origin. But ecological effects on wild or weedy plant species may be significant as well--both those favored (as in the spontaneous native "crops" of the Lacandon Maya) and those considered noxious or detrimental to agriculture. The latter group--weeds and ruderals, are likely to undergo rapid selection favoring genotypes that outsmart the farmer (Anderson 1967). Likewise, insect pests, though slowed by the complexity of mixed-species agriculture (Altieri 1991), are likely to benefit from the presence of high concentrations of crops.

Botanical evidence from present-day forests, reports of early Hispanic visitors, and inferences from contemporary practices of Mayan descendants, such as the Lacandon, suggests that the ancient Olmec and Mayan civilizations of southern Mexico and Guatemala had developed sophisticated systems of agroforestry. Many kinds of useful, native tree species were either spared as spontaneous seedlings or planted in polycultural plots. Similar systems appear to have been common in other tropical regions. While these systems certainly favored some plant species (and their associates) over others, in southern Mexico, neither the nature of the habitat nor the overall composition of the extremely rich, regional flora appears to have been greatly affected. Moreover, at the height of Maya culture, these systems supported some 400-500 people per square km in rural areas, on a sustainable basis--compared with about 5 in the same habitats today (Gomez-Pompa and Kaus 1990).

In both ecological and evolutionary terms, the botanical aspect of the Mayan component community parallels rather well, perhaps, that of the leaf-cutter ants. The ants are dependent on one, and the Maya relied on a few highly-modified cultivars, with significant but sustainable effects on a wide spectrum of wild plant species.

The present dilemma

In contrast with the relatively circumscribed domain of human interactions with other species in ancient times, the history of modern society tells a tale of increasingly strong interactions, both evolutionary and ecological, with an accelerating number of species. The human "component" community now extends to species in virtually every corner of every ecosystem on earth. Now, we are indeed on the threshold of an age in which our actions may affect, literally, all species on the planet though global climate change.

Massive habitat alterations for agriculture and settlements are the most obvious cause of the extension of our influence on other species. In some regions, such as the Mediterranean Basin, this process begun thousands of years ago. But, ever more frequently, with habitat alteration, human influence on species that were not traditionally part of our component community reaches the limit--in their extinction, which qualifies as both an ecological and evolutionary event. The effects of an extinction on other species in a community or on ecosystem function may range from negligible to catastrophic, but the loss of the extinct species as a potential ancestor and the loss of its potential for human utility are absolute and permanent. The alarming current and projected rate of species extinctions has been frequently documented and discussed (e.g. Myers 1979, Ehrlich and Ehrlich 1981, Wilson 1988).

Plantings of genetic monocultures, often covering millions of hectares, have eroded not only the genetic diversity of most major food crops (Keystone Center 1991), but have actually reduced the spectrum of cultivated species actively grown in agriculture especially in tropical areas (National Academy of Sciences 1975). Recent interest in reviving such crops as quinoa (*Chenopodium quinoa*), an ancient Andean crop that is more nutritious than any cereal grain (Wood 1988), and remains a staple among traditional Andean cultures, has helped rescue some species from potential neglect and extinction in the face of modern agriculture.

The introduction of exotic species of plants, animals, and microorganisms, both intentional and accidental, has had the effect of homogenizing the biota of the earth. This spreading homogeneity results not only from the expanding geographic range of pest species, but from the displacement, range reduction, or even extinction of native species everywhere. They are the victims of competition, predation, or the diseases of the interlopers (Hughes 1986; Singer *et al.* 1984; Mooney and Drake 1986; Sinderman *et al.* 1991).

In many species, rapid evolutionary changes resulting from human activities are well documented. These changes have nothing to do, however, with the intentional domestication of new species; not a single significant new domesticate is known to have been produced for millenia (Anderson 1967), although immense efforts have been directed at the genetic improvement of existing crops and livestock. The changes I refer to are unintended ones, such as the evolution of industrial melanism (Kettlewell 1973), antibiotic resistance, heavy metal resistance, and pesticide resistance. In the case of insecticides, not only tens of thousands of species of intended insect victims, but even unintended targets have evolved resistance--there are insecticide-resistant rodents and fish, for example. In the case of heavy-metal tolerance, which is known from plants on mine dumps and from soil and aquatic microbes, no victim was ever intended.

While these genetic changes testify to the adaptive potential of organisms (short lived ones) in the novel chemical environments we have created, they are also an indictment of the pervasive and continuing human alteration of the biosphere. The impending threat of global change in climate and in biogeochemical processes from progressive alterations in atmospheric chemistry pose not only an ecological but an evolutionary challenge to every species on the planet. Surely many will survive the ecological changes, and some short-lived species will even adapt to them genetically. What is not clear is that this coming new world is one in which we can or would want to live in.

The loss of genetic resources and decline of traditional knowledge

In parallel with the spread of introduced species and mechanized agriculture, the elimination or transformation of traditional societies has already decimated, in many cases, the unwritten library of agricultural and silvicultural wisdom that these cultures developed over millennia by trial and error (Denslow and Padoch 1988). Because of the idiosyncratic nature of species interactions--for example, among crop and tree species grown in polyculture, and between pests or beneficial animals and crops--traditional knowledge of this kind tends to be highly local in its applicability. Thus, even resettlement of supposedly intact traditional societies is likely to reduce the effectiveness of finely tuned agricultural practices and the yield of the land races of their crops.

The important phenomenon called "over-yielding" in polycultures (Altieri 1991) refers to higher productivity in mixed-species plantings (e.g. maize-beans-squash) than would be predicted from values based on pure stands of the same crops. Over-yielding very likely depends in part on the adaptation of the component crops to each other, under local soil conditions and cultural practices. In effect, overyielding itself is a property selected for by the farmer at the level of the polyculture component community (see

Wilson [1980], for theoretical background). Thus, substitution of commercial varieties for traditional ones in polycultures is likely to eliminate the value and obscure the function of the biotic adaptations of land races to one another.

Like the adaptations of wild species, the adaptations of local crops and land races have, under human cultivation, been selected to survive the effects of relatively rare climatic events or pest outbreaks that may occur only once a decade or even once a century. In contrast, exotic species or higher-yielding commercial varieties imported from elsewhere may do well in a new site for a few years, replacing local crops or land races, but then undergo massive failure in the face of such a rare event.

Efforts to record and to learn from the practices of traditional cultures, to protect them from unwanted transformation, and to integrate traditional knowledge with contemporary agricultural practices must be an integral part of any program aimed at preserving or restoring biodiversity. In many cases, it is already too late, but rising interest in this objective is encouraging (Denslow and Padoch 1988; Gomez-Pompa and Kaus 1990; Altieri 1991).

Economies, ecology and ethics

In our ancestral component community, economic and ecological relations with biological resources were one and the same. Even as market economies began to develop, economic and ecological relations were roughly commensurate in their geographic domain. In contrast, urban industrial society in the present day trades in biological resources on a global scale, while the ecological effects have, until recently, been presumed to remain at the same local scale as they did in the Paleolithic. In short, international resource economics has been uncoupled, for the past century or two, from local resource ecology. The headlong harvest of wild species and permanent alteration of wildlands and waters for short-term economic gain has often left long-term ecological and consequent economic costs to be reckoned by others. Often, those others are the least able to bear these costs--poor farmers in third world countries.

The field of environmental ethical philosophy, though still in its infancy by philosophical standards, has made progress with the fundamental questions that arise from the philosophical heresy of a non-anthropocentric ethics (Mannison et al. 1980; Norton 1986, 1987; Taylor 1986; Ehrenfeld 1988; Callicott 1989; Sitter 1989; McNeely, 1990). A principal issue in this field is the question of the value of natural entities--species and ecosystems, in particular.

One of the many taxonomies of value makes a first distinction between instrumental vs. intrinsic value. Instrumental value refers to all varieties of actual or potential usefulness to humans. Intrinsic value is a quality that, by definition, lies beyond any utility. For example, in our own culture and its embodiment in law, each human being has intrinsic value, regardless of his or her condition or usefulness to society. Arguments for the intrinsic value of species are complex and controversial. I have discussed them at length elsewhere, from an evolutionary ecologist's point of view, along with a review of some aspects of instrumental value (Colwell 1989). Here I will recapitulate the discussion of the latter topic.

The instrumental value of genotypes, species, and ecosystems, as a practical argument, boils down to economics (see McNeely *et al.* [1990], who use a somewhat different taxonomy of values). What is the economic cost of declining biodiversity, of habitat loss and species extinctions?

Genetic resources and biochemical prospecting

The economic argument for the preservation of wild relatives of key domesticated species has always been clear enough: they represent a potential source of commercially useful genetic material--"germplasm" (Witt 1985; Williams 1988; Keystone Center 1991). In the past century, scientific breeders of plants and animals have reached back into the evolutionary history of domesticated species to recapture useful genetic traits from their wild relatives--sometimes from the true ancestral species, sometimes from evolutionary cousins. Resistance to disease, pests, or stress; nutrient balance; growth form; and fruit shape or quality have been developed in crops through hybridization with wild relatives, followed by complex breeding programs to combine desired traits in a single strain (Goodman *et al.* 1987; Iltes 1988). Improved breeds of domesticated animals have sometimes been developed in the same way. The molecular and cellular techniques of biotechnology (of which more will be said later) have already begun to speed this process immensely.

But what about the multitude of wild species--the vast majority of both the plant and animal kingdoms--that have no domesticated or commercially valuable relatives? At least two general classes of economic arguments have been advanced for the preservation of these "unexploited" species. First, some of these species, in themselves, may prove to be of direct commodity value (Norton 1988) or productive use value (McNeely *et al.* 1990) in the human enterprise, or alternatively, may be valued for their potential commercial usefulness at some future time (an aspect of option value) (Randall 1988). The second class of economic valuations rests on attempts to estimate the non-commodity (or non-market or amenity) value of species, or of biodiversity itself, as measured by the degree to which people consider the economic value of places, services, or experiences to be increased by the presence or existence of species or by biodiversity (Kellert 1986; Norton 1987, 1988; Randall 1988).

Few would base the argument for the commodity value of currently unexploited species on the prospect that important, totally new food plants are likely to be discovered among wild species--although many known food plant species with excellent nutritional value and promising ecological characteristics are underutilized (National Academy of Sciences 1975; Vietmeyer 1986; Wood 1988). Prospects for the discovery of novel biochemical compounds, however, have motivated several well-funded, intensive commercial surveys of pharmacologically or biologically active natural plant and animal products for possible use as drugs, biocides, and industrial biochemicals (Eisner 1990). Some of these efforts focus on plants used in the practice of folk medicine in traditional cultures, whereas others are simply broad surveys of plant (and in one case in Costa Rica, insect) material collected more or less at random, especially in tropical forests (Lewis and Elvin-Lewis 1977; Myers 1983; Farnsworth 1988). There are concerns, however, that the advent of computer-designed molecules and the techniques of genetic engineering may displace these efforts in the pharmaceutical industry (Ehrenfeld 1988).

Biotechnology and biodiversity

Crosses between species more distantly related than members of very closely-related genera were impossible for higher organisms until the past decade, when the molecular and cellular techniques of biotechnology became feasible. Now, functional genetic units, and the phenotypic traits they create or modify, are being moved not only between closely related species, but between kingdoms. One typical example among a burgeoning number: the first genetically engineered plant to be approved for field-testing in the U.S.A. (approved in 1985) was a herbicide-tolerant tobacco strain constructed using genetic material from a bacterium (a *Salmonella* species that had become resistant to the herbicide), controlled by additional genetic sequences from a mammal (sheep) and another plant (soybean) unrelated to tobacco, all inserted using a second species of bacterium (*Agrobacterium*) (Comai *et al*. 1983).

Some have suggested that the tools of biotechnology will actually add to the pool of biological diversity, by creating new genetic combinations not possible or not likely in nature, or even by the addition of completely synthetic genes to the products of evolution. In whatever measure this ambitious prediction comes true, the implication that laboratory art can truly imitate life betrays a narrowly reductionist view of adaptation and evolution. The diversity of organisms in nature arises from the interplay of genetic variation with the exigencies of life in environments teeming with other organisms and buffeted by shifting physical factors. The adaptive "inventions" of natural selection seem far more likely to be of use in creating new products and, with luck and skill, solving serious ecological problems than any biological feature produced *de novo*. After all, nature has a head start on us of many hundreds of millions of years and maintains a hundred million natural laboratories operating 24 hours a day.

Gene technologies clearly stand among the ultimate beneficiaries of the vast library of tried-and-true evolutionary inventions of the millions of species in natural ecosystems and thus have an interest in keeping these libraries in viable condition (Goodman *et al*. 1987; Janzen 1987; Witt 1985). The difficulties in transferring any but the simplest traits to an unrelated species are currently formidable, but there is every reason to expect that many difficulties will be overcome rapidly. Twenty years ago, most biologists would have declared impossible--or at least extraordinarily unlikely--what has already been accomplished today.

Non-commodity value of species and ecosystems

Economic arguments for the preservation of species (or of biodiversity or habitat) based on non-commodity values rely on measures of the degree to which the value of a place is enhanced by the presence, or decreased by the loss, of particular species (or of a habitat). The value differential may be estimated from actual prices, for example by comparing the sale price of architecturally equivalent homes at increasing distances from a nature preserve. Alternatively, survey methods may be used to assess how much citizens would, in principle, be willing to pay to keep a species or habitat or how much they would be willing to accept as compensation for its loss. In theory, such surveys could be used to estimate the value people in one part the world place on the very existence of a particular species or group of species elsewhere, which they may never have seen and will never see, outside photographs or films (for example, the blue whale) (McNeely *et al*. 1990; Randall 1988).

However important non-commodity valuation may be in particular cases (Stoll and Johnson 1984), an exclusively species-by-species approach to the non-commodity economic valuation of biodiversity is not only impractical and liable to underestimate consistently the "contributory value" of species to ecosystem function (Norton 1987), but cannot be expected to lead to the even-handed protection of food webs and ecosystems--the key to long-term preservation. The strategy of protecting entire habitats and ecosystems by focusing popular appeals on familiar and evocative species, as exemplified by the campaign to save the giant panda, strikes many biologists as manipulative and somewhat disingenuous--but, nonetheless, most accept it as a means to an end. If the "amenity value" and "existence value" of natural entities can be increased by educational efforts (Randall 1986), then the best program for the promotion of even-handed species preservation from this strictly economic viewpoint calls not only for an attempt to get people to love tapeworms, termites, and toads, but in addition a full scale effort to give biodiversity itself the same cachet as pandas and penguins.

Conclusion

Many ecologists share the conviction that a full accounting of the value of intact ecosystems would surely show, unequivocally, that investment in sustaining biological resources is the best economic option, quite aside from ethical considerations. The indirect economic value of ecosystem services--regulation of climate, recycling of nutrients and gases, purification of groundwater, decomposition of organic matter--though virtually impossible to calculate, at least is beginning to be more generally appreciated (McNeely *et al*. 1990) We still do not know very much, however, about the role of biodiversity in ecosystem function, beyond the obvious limiting cases. The role of particular species, in particular cases, may be either crucial or negligible. I have attempted to show that other reasons for the preservation of species, habitats, and biological communities may be forcefully argued.

The future value of biodiversity (and of the diversity of human knowledge of traditional uses of wild species) can only increase, as the worldwide rate of extinction accelerates. As long as the perverse alchemy of non-sustainable exploitation of biological resources continues to turn gold into lead; those who sustain the diversity of living things and their genetic inventions will find themselves custodians of a treasure trove of ever-increasing worth.

References

Altieri, M. A. 1991. Ecology of tropical herbivores in polycultural agroecosystems. In P. W. Price, T. Lewinsohn, G. W. Fernandes and W. W. Benson (eds.), *Plant-animal interactions: Evolutionary ecology in tropical and temperate regions*, pp. 607-617. New York: John Wiley and Sons.

Anderson, E. 1967. *Plants, man, and life*. Berkeley, Ca.: University of California Press.

Baker, H. G. 1978. Chemical aspects of the pollination biology of woody plants in the tropics. In P. B. Tomlinson and M. G. Zimmerman (eds.), *Tropical trees as living systems*, pp. 57-82. New York: Cambridge University Press.

Boucher, D. H., S. James, and K. S. Keeler. 1982. The ecology of mutualism. *Annual Review of Ecology and Systematics* 13:315-347.

Callicott, J. B. 1989. *In defense of the land ethic*. Albany, NY.: State University of New York Press.

Colwell, R. K. 1985. The evolution of ecology. *American Zoologist* 25: 771-777.

Colwell, R. K. 1986a. Community biology and sexual selection: Lessons from hummingbird flower mites. In T. J. Case and J. Diamond (eds.), *Ecological communities*, pp. 406-424. New York: Harper and Row.

Colwell, R. K. 1986b. Population structure and sexual selection for host fidelity in the speciation of hummingbird flower mites. In S. Karlin and E. Nevo (eds.), *Evolutionary processes and theory*, pp. 475-495. New York: Academic Press.

Colwell, R. K. 1989. Natural and unnatural history: Biological diversity and genetic engineering. In W. R. Shea and B. Sitter (eds.), *Scientists and their responsibility*, pp. 1-40. Canton, MA: Watson International Publishing.

Comai, L., L. C. Sen, and D. M. Stalker. 1983. An altered aroA gene product confers resistance to the herbicide glyphosate. *Science* 221: 370-371.

Denslow, J. S., and C. Padoch. 1988. *People of the tropical rain forest*. Berkeley, CA: University of California Press.

Dobley, J. 1990. Molecular evidence for gene flow among *Zea* species. *Bioscience* 40: 443-448.

Ehrenfeld, D. 1988. Why put a value on biodiversity? In E. O. Wilson (ed.), *Biodiversity*, pp. 212-216. Washington, D. C.: National Academy Press.

Ehrlich, P. R., and A. Ehrlich. 1981. *Extinction: The causes and consequences of the disappearance of species*. New York: Random House.

Ehrlich, P. R., and P. H. Raven. 1964. Butterflies and plants: A study in coevolution. *Evolution* 18: 586-608.

Eisner, T. 1990. Prospecting for nature's chemical riches. *Issues in Science and Technology* 6 (No.2: winter 1989-1990): 31-34.

Farnsworth, N. R. 1988. Screening plants for new medicines. in E. O. Wilson (ed.), *BioDiversity*, pp. 83-97. Washington, D. C.: National Academy Press.

Feeny, P. 1991. Chemical constraints on the evolution of swallowtail butterflies. In P. W. Price, T. Lewinsohn, G. W. Fernandes and W. W. Benson., (eds.), *Plant-animal interactions: Evolutionary ecology in tropical and temperate regions*, pp. 315-340. New York: John Wiley and Sons.

Feinsinger, P. 1987. Approaches to nectarivore-plant interactions in the New World. *Revista Chilena de Historia Natural* 60: 285-319.

Feinsinger, P., R. K. Colwell, J. Terborgh, and S. B. Chaplin. 1979. Elevation and the morphology, flight energetics, and foraging ecology of tropical hummingbirds. *American Naturalist* 113: 481-497.

Gardner, M. R., and W. R. Ashby. 1970. Connectance of large dynamic (cybernetic) systems: critical values for stability. *Nature* 228:784.

Gilbert, L. E., and J. H. Smiley. 1978. Determinants of local diversity in phytophagous insects: host specialists in tropical environments. In A. Mound and N. Waloff, (eds.), *Diversity of insect faunas*, pp. 89-105. Royal Entomological Society of London, Symposium 9.

Gomez-Pompa, A., and A. Kaus. 1990. Traditional management of tropical forests in Mexico. In A. B. Anderson (eds.), *Alternatives to deforestation: Steps toward sustainable use of the Amazon rain forest*, pp. 45-64. New York: Columbia University Press.

Goodman, R. M., H. Hauptli, A. Crossway, and V. C. Knauf. 1987. Gene transfer in crop improvement. *Science* 236:48-54.

Hames, R. B., and W. T. Vickers (ed.), 1983. *Adaptive responses of Native Amazonians*. New York: Academic Press.

Howard, J. J. 1988. Leafcutting ant diet selection. Relative influence of leaf chemistry and physical features. *Ecology* 69: 250-260.

Hughes, N. F. 1986. Changes in the feeding biology of the Nile perch, *Lates niloticus* (L.) (Pisces: Centropomidae), in Lake Victoria, East Africa, since its introduction in 1960, and its impact on the native fish community of the Nyanza Gulf. *Journal of Fisheries Biology* 29: 541-548.

Iltes, H. H. 1988. Serendipity in the exploration of biodiversity: What good are weedy tomatoes? In E. O. Wilson (ed.), *BioDiversity*, pp. 98-105. Washington, D. C.: National Academy Press.

Janzen, D. H. 1987. Conservation and agricultural economics. *Science* 236:1159.

Kellert, S. R. 1986. Social and perceptual factors in the preservation of animal species. In B. G. Norton (ed.), *The preservation of species*, pp. 50-73. Princeton, N. J.: Princeton University Press.

Kettlewell, H. B. D. 1973. *The evolution of melanism*. Oxford, U.K.: Clarendon Press.

Keystone Center. 1991. Final consensus report: Global initiative for the security and sustainable use of plant genetic resources. *Keystone Dialogue Series on Plant Genetic Resources.* Keystone, Colo.: Keystone Center.

Lewis, W. H., and M. P. F. Elvin-Lewis. 1977. *Medical botany*. New York: John Wiley and Sons.

Mannison, D., M. McRobbie, and R. Routley, (eds.) 1980. *Environmental Philosophy*. Canberra: Australian National University.

May, R. M., ed. 1981. *Theoretical ecology*, 2nd ed. Oxford, U.K.: Blackwell Scientific Publications.

McNeely, J. A., K. R. Miller, W. V. Reid, R. A. Mittermeier, and T. B. Werner. 1990. *Conserving the world's biological diversity.* Gland, Switzerland: International Union for the Conservation of Nature.

Mooney, H. A., and J. A. Drake (eds.), 1986. *Ecology of biological invasions of North America and Hawaii*. New York: Springer-Verlag.

Myers, N. 1979. *The Sinking ark*. Oxford, U. K.: Pergamon Press.

Myers, N. 1983. *A wealth of wild species: Storehouse for human welfare*. Boulder, Colorado: Westview Press.

National Academy of Sciences, U. S. A. 1975. *Underexploited tropical plants with promising economic value*. Washington, DC.: National Academy of Sciences Press.

Nations, J. D. 1988. The Lacandon Maya. In J. S. Denslow and C. Padoch, (eds.), *People of the tropical rain forest*, pp. 86-88. Berkeley, Ca.: University of California Press.

Nations, J. D., and R. B. Nigh. 1980. The evolutionary potential of the Lacandon Maya sustained-yield tropical forest agriculture. *Journal of Anthropological Research* 36: 1-30.

Newstrom, L. E., G. W. Frankie, H. G. Baker, and R. K. Colwell. (In press). Flowering phenology at La Selva. In K. S. Bawa, L. McDade and G. S. Hartshorn, (eds.), *La Selva: ecology and natural history of a lowland tropical rain forest*. Chicago: University of Chicago Press.

Norton, B. G. 1986. On the inherent danger of undervaluing species. In Norton, B. G. (ed.), *The preservation of species,* pp. 110-137. Princeton, N. J.: Princeton University Press.

Norton, B. G. 1987. *Why preserve natural variety?* Princeton, N. J.: Princeton University Press.

Norton, B. G. 1988. Commodity, amenity, and morality: the limits of quantification in valuing biodiversity. In Wilson, E. O. (ed.), *BioDiversity*, pp. 200-205. Washington, D. C.: National Academy Press.

Price, P. W., T. Lewinsohn, G. W. Fernandes, and W. W. Benson, (eds.) 1991. *Plant-animal interactions: Evolutionary ecology in tropical and temperate regions*. New York: John Wiley and Sons.

Quinlan, R. J. and J. M. Cherrett. 1979. The role of fungus in the diet of the leaf-cutting ant *Atta cephalotes* (L.). *Ecological Entomology* 4: 151-160.

Randall, A. 1986. Human preferences, economics, and the preservation of species. In B. G. Norton (ed.), *The preservation of species*, pp. 79-109. Princeton, N. J.: Princeton University Press.

Randall, A. 1988. What mainstream economists have to say about the value of biodiversity. In Wilson, E. O. (ed.), *BioDiversity*, pp. 217-223 Washington, D. C.: National Academy Press.

Root, R. B. 1973. Organization of a plant-arthropod association in simple and diverse habitats: the fauna of collards (*Brassica oleracea*). *Ecological Monographs* 43: 95-124.

Simpson, B. B., and M. Conner-Ogorzaly. 1986. *Economic botany: Plants in our world*. New York: McGraw Hill.

Sinderman, C. J., B. Steinmetz, and W. K. Hershberger, (ed.) 1991. *Effects of introduction and transfers of aquatic species on resources and ecosystems*.

Singer, F. J., W. T. Swank, and E. E. C. Clebsch. 1984. Effects of wild pig rooting in a deciduous forest. *J. of Wildlife Management* 48: 464-473.

Sitter, B. 1989. In defence of non-anthropocentrism in environmental ethics. In W. R. Shea and B. Sitter (eds.), *Scientists and their responsability*, pp. 105-145. Canton, Ma.: Watson International Publishing.

Skutch, A. 1973. *The life of the hummingbird.* New York: Vineyard Books.

Stevens, G. C. 1983. *Atta cephalotes* (zompopas, leaf-cutting ants). In D. H. Janzen, (ed.), *Costa Rican natural history,* pp. 688-691. Chicago: University of Chicago Press.

Stoll, J., and L. Johnson. 1984. Concepts of value, nonmarket valuation, and the case of the whooping crane. *Transactions of the North American Wildlife Natural Resource Conference* 49: 382-393.

Taylor, P. W. 1986. *Respect for nature.* Princeton, N. J.: Princeton University Press.

Vietmeyer, N. D. 1986. Lesser-known plants of potential use in agriculture and forestry. *Science* 232:1379-1384.

Williams, J. T. 1988. Identifying and protecting the origins of our food plants. In E. O. Wilson (ed.), *BioDiversity*, pp. 240-247. Washington, D. C.: National Academy Press.

Wilson, D. S. 1980. *The natural selection of populations and communities.* Menlo Park, Ca: Benjamin Cummings.

Wilson, E. O. 1980. Caste and division of labor in leaf-cutter ants (Hymenoptera: Formicidae: *Atta*). II. The ergonomic optimization of leaf cutting. *Behavioral Ecology and Sociobiology* 7: 157-165.

Wilson, E. O. (ed.) 1988. *BioDiversity.* Washington, D. C.: National Academy Press.

Witt, S. C. 1985. *Biotechnology and genetic diversity.* San Francisco, Ca.: California Agricultural Lands Project.

Wood, R. 1988. *Quinoa, the supergrain: Ancient food for today.* New York: Harper and Row.

Index